A.C. Machines:
Electromagnetics and Design

ELECTRONIC & ELECTRICAL ENGINEERING RESEARCH STUDIES

ELECTRICAL MACHINES SERIES

Series Editor: **Dr Peter J. Tavner**
Lintott Control Equipment plc, England

1. Condition Monitoring of Electrical Machines
   Peter J. Tavner and James Penman

2. Response Analysis of A.C. Electrical Machines: Computer Models and Simulation
   John R. Smith

3. A.C. Generators: Design and Application
   Robert L. Ames

4. A.C. Machines: Electromagnetics and Design*
   Brian Chalmers *and* **Alan Williamson**

*** Cover Photograph**
Rotor of a 1485 MW, 20 kV, 4-pole turbine generator
Courtesy of GEC Alsthom Turbine Generators Limited

N. JENKINS

A.C. Machines: Electromagnetics and Design

Brian Chalmers *and* **Alan Williamson**
UMIST, UK

RESEARCH STUDIES PRESS LTD.
Taunton, Somerset, England

JOHN WILEY & SONS INC.
New York · Chichester · Toronto · Brisbane · Singapore

RESEARCH STUDIES PRESS LTD.
24 Belvedere Road, Taunton, Somerset, England TA1 1HD

Marketing and Distribution:

Australia and New Zealand:
JACARANDA WILEY LTD.
GPO Box 859, Brisbane, Queensland 4001, Australia

Canada:
JOHN WILEY & SONS CANADA LIMITED
22 Worcester Road, Rexdale, Ontario, Canada

Europe, Africa, Middle East and Japan:
JOHN WILEY & SONS LIMITED
Baffins Lane, Chichester, West Sussex, England

North and South America:
JOHN WILEY & SONS INC.
605 Third Avenue, New York, NY 10158, USA

South East Asia:
JOHN WILEY & SONS (SEA) PTE LTD
37 Jalan Pemimpin #05-04
Block B Union Industrial Building, Singapore 2057

Library of Congress Cataloging in Publication Data

Chalmers, B. J. (Brian John)
A.C. machines : electromagnetics and design / Brian Chalmers and Alan Williamson.
p. cm. — (Electronic & electrical engineering research studies. Electrical machines series ; 4)
Includes bibliographical references and index.
ISBN 0-86380-115-3. — ISBN 0-471-93003-2 (Wiley)
1. Electric machinery—Alternating current—Design and construction. 2. Electromagnetism. I. Williamson, A. C. (Alan Charles), 1935– . II. Title. III. Series.
TK2725.C48 1991
621.31'33--dc20 90-27686
CIP

British Library Cataloguing in Publication Data

Chalmers, Brian John
A.C. machines: Electromagnetics and design.
– (Electronic and electrical engineering research studies – electrical machines series)
I. Title II. Williamson, Alan C. III. Series
621.31

ISBN 0 86380 115 3

ISBN 0 86380 115 3 (Research Studies Press Ltd.)
ISBN 0 471 93003 2 (John Wiley & Sons Inc.)

Printed in Great Britain by SRP Ltd., Exeter

Editorial Foreword

The Electrical Machines Series is designed to provide books of immediate use to practising engineers and research workers.

Brian Chalmers and Alan Williamson have had many years' experience of electrical machines as research workers, in the design office and as undergraduate and postgraduate teachers. This has equipped them well to provide the clear exposition which is presented here. They have particular experience of the electromagnetic problems of electrical machine design and their published analytical work is well known to the international machines community.

This book provides a thorough description of the electromagnetic design of electrical machines, complete with relevant equations and a bibliography. It will be an essential guide to machine design for manufacturers, users and researchers alike.

Dr. P.J. Tavner
Engineering Director
Lintott Control Equipment PLC

Preface

Our early periods of involvement in the development and design departments of the English Electric Company at Stafford gave very valuable exposure to the industrial world of electrical machines. We benefitted from close involvement with designers who had many years of experience, while our specific role was to tackle special problems which lay outside the normal routine. The machine types encompassed turbine generators, induction motors and salient-pole motors and generators.

Subsequent years in the academic environment at UMIST have given the opportunity to develop parallel threads of activity in teaching and research. The desire to explain things clearly to undergraduate students, and to attempt to instill some enthusiasm for the subject, has continuously exercised our minds. At the same time, the urge to pursue original and industrially-relevant research in a university department with a long tradition at postgraduate level has led us down paths of discovery and clarification.

The contents of this book are a distillation of some of the fruits of these experiences. First produced as notes for a short postgraduate course and subsequently delivered to designers at the site of a major manufacturer, it is hoped that this publication will be particularly timely when there is a need to recruit new designers and to equip them for their task. Its contents are, in essence, a presentation of established analytical methods, reinforced by personal treatments of topics such as winding analysis, harmonics, steady-state and transient models, eddy currents and an overview of scale effects.

It may be thought that there is a somewhat old-fashioned look about some sections of the book, for which we offer no apology. We do not claim to present an up-to-date review of machine design procedures. Increasing use is being made of numerical methods for solution of magnetic fields to compute various aspects of machine behaviour or properties, such as inductances, and some machine models use equivalent circuits which are more complex than those treated in this book. However, it is our belief that these procedures do not help obtain insight into either the more significant facets of machine operation or the influence of major design features upon performance. Any numerical solution is specific to the particular case, whereas an analytical treatment describes quite general trends of variation. Any section of the analysis may, of course, be replaced by a more detailed numerical computation, if desired.

It is a pleasure to acknowledge with gratitude the contribution to the origins of this book made by former colleagues, particularly Noel Adcock, Dennis Plevin, Albert Reece and Mike Tarkanyi.

B.J. Chalmers
A.C. Williamson
Manchester
November 1990

Contents

Principal Symbols

B	flux density
D	electric displacement, diameter
d	depth
E	electric field strength, voltage
e	voltage
F	magnetomotive force (mmf), force
f	frequency
g	air-gap length
H	magnetic field strength
I	current
i	current
J	current density
K_w	winding factor = K_pK_d
L	core length, conductor length, inductance
l	leakage inductance
M	mutual inductance
m	number of phases
N	number of turns (in series per phase)
n	rotational speed, harmonic order
P	permeance, power
p	number of pole-pairs
q	slots per pole per phase
R	resistance
r	resistance

S	number of slots
s	slots per pole, fractional slip
T	torque, time constant
t	time
U	stored energy
V	voltage
v	voltage, velocity
w	width
X	general reactance
x	leakage reactance
Z	impedance
δ	depth of penetration
ϵ	permittivity = $\epsilon_r \epsilon_o$
θ	angle
λ	flux linkage, specific permeance coefficient
μ	permeability = $\mu_r \mu_o$
ρ	resistivity
τ	pole pitch
Φ	flux
φ	phase angle
ω	angular frequency

Suffices

1,2	stator, rotor
b	rotor bar
c	core, cylindrical rotor
d,q	direct, quadrature axes
e	electromagnetic, effective, electrical
f	field
m	magnetising, mechanical
n	harmonic
t	teeth
s,r	stator, rotor

D,Q direct, quadrature axis damper

Vector quantities are in bold type (e.g. **B**)
Complex and phasor quantities are shown thus: $\bar{Z}$ and $\bar{I}$

CHAPTER 1
Electromagnetic Relationships

1.1 Introduction

It is assumed that the reader has some acquaintance with electromagnetism, being familiar with the concept of a "field" and aware of the definitions of basic quantities, and has some knowledge of the conventional geometries of the main types of machine and the essential differences between them.

To aid familiarity, the symbols used in this book are, as far as possible, those commonly used by electrical engineers involved in machine design and analysis.

Machine design is not an exact science and many approximations are made in order to make a design process practicable. Fortunately, conventional machine geometry permits most of the approximations to be made without serious error, enabling, for example, certain end effects to be considered independently and certain interactions between fields to be ignored, as in the calculation of leakage inductance.

The material contained in this Chapter is of a basic nature and is designed to give an insight into the mechanisms by which an electrical machine develops power and the techniques and approximations used in the electromagnetic aspects of machine design.

1.2 Maxwell's Equations

Maxwell brought together a unifying set of equations which relate all electric and magnetic field quantities. These equations are usually written in vector notation and are, with no significance in order,

$$\text{div}\ \mathbf{D} = \sigma \qquad 1.1$$
$$\text{div}\ \mathbf{B} = 0 \qquad 1.2$$
$$\text{curl}\ \mathbf{H} = \mathbf{J} + \partial\mathbf{D}/\partial t \qquad 1.3$$
$$\text{curl}\ \mathbf{E} = -\partial\mathbf{B}/\partial t \qquad 1.4$$

with $\mathbf{D} = \varepsilon\mathbf{E}$,

$\mathbf{B} = \mu\mathbf{H}$

and $\mathbf{E} = \rho\mathbf{J}$

1.3 Electromagnetics at Power Frequencies

Equation 1.1 is of interest in the study of electric fields, an aspect of machines which, although particularly important to insulation engineers, is not covered in this book. We are concerned essentially with electromagnetics.

It is very convenient to use the concept of a line of magnetic flux when dealing with magnetic circuits. The fact that we can use this concept is implied by equation 1.2, which states that the net magnetic flux entering any volume is zero. Consequently a magnetic flux line is continuous; there are no "sources" of magnetic flux.

In magnetic circuits linked by current-carrying conductors at low frequencies (less than about 10^{16}Hz!) the displacement current ($\partial\mathbf{D}/\partial t$) can be neglected compared with conduction current ($\mathbf{J}$) and we shall ignore it. Equation 1.3 can be written in the form of the familiar circuital law of Ampère. This states that the line integral of **H** around <u>any</u> closed path is proportional to the total current flowing through the loop formed by the path of integration. In SI units this can be written as

$$\oint \mathbf{H}.d\mathbf{l} = \sum i \qquad 1.5$$

where $\sum i$ is the sum of conductor currents linked by the integration path. We shall use the name magnetomotive force, or mmf, for the value of this integral. Note that mmf is a scalar quantity, i.e. not a vector.

Equation 1.4 also has a more familiar form, which is known as Faraday's Law of Induction. This states that, whenever the magnetic flux linking an electric circuit changes, a voltage is induced around the circuit proportional to the rate of change of flux linkage.

In SI units this can be written as

$$e = d\lambda/dt \qquad 1.6$$

where λ is the flux linkage. λ is often written as $N\Phi$ where N is the number of turns of a coil and Φ is the effective flux linking all N turns.

A negative sign is sometimes included in equation 1.6 to stress the fact that the direction of e is always such that, if the circuit is closed, a current will flow to produce a flux opposing the change. This is known as Lenz's Law. However, it is considered that the negative sign can cause more confusion than clarification, since a little thought shows that conservation of energy demands that e be a voltage opposing the change.

Faraday's Law applies for all cases, and includes that of motional emf, where λ changes as a result of a change in the relative position of a circuit, as well as transformer emf, where λ changes as a result of changing currents or mmf.

1.4 Magnetic Circuits

The determination of B at every point in a region containing irregular-shaped masses of magnetic material is a boundary-value field problem which is very difficult to solve rigorously. The difficulty is associated with the usually complex nature of the boundaries and with the non-linear relationship between applied field strength and flux density for ferro-magnetic materials. To solve precisely a field problem in iron it is necessary to have a knowledge of the previous history of the material in order to determine where on the characteristic the present working point is situated. This arises because of hysteresis shown by the dotted line in the typical characteristics shown in Fig. 1.1. In all but permanent magnet machines it is usual to ignore hysteresis when solving for flux distributions, and to assume the characteristic to be similar to that shown by the full-line of Fig. 1.1. This is usually satisfactory because low-hysteresis steels are often used.

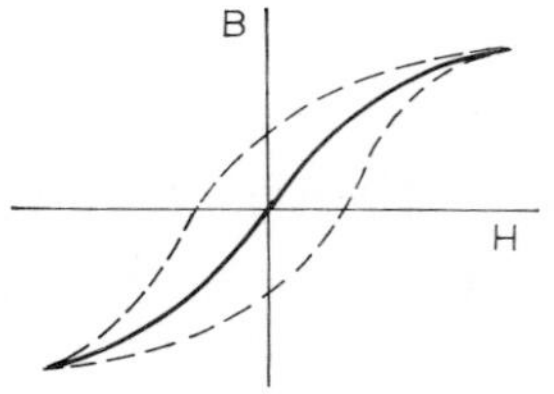

Fig. 1.1.
B-H characteristic of
ferromagnetic material

In the area of electrical machine design approximations are frequently made, to render magnetic circuits less daunting. To illustrate the type of approximation made, a "simple" arrangement is considered.

Fig. 1.2 shows the section through a magnetic core containing an airgap and wound with a coil. A very crude approximation is to assume a flux distribution as indicated by the dotted flux lines. Note that such an assumption implies that the circuit has already been analysed. However, even in quite complex circuits it is possible with experience to predict the principal flux directions.

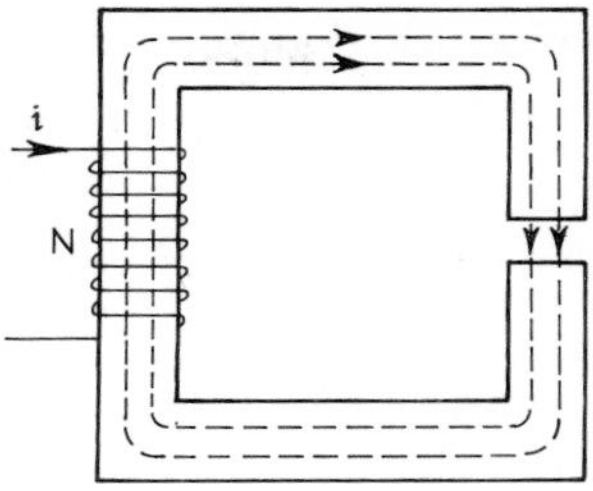

Fig. 1.2.
A simple magnetic circuit

The sketch assumes that :-

(i) the only flux flowing in the iron is that which crosses the airgap

(ii) flux density is uniform across the airgap and over its area of cross-section

(iii) there is no fringing at the edges of the gap

(iv) flux density is uniform over the cross-section at any point around the magnetic circuit

(v) the pattern of flux at the limb corners is idealised as shown.

In order to determine the relationship between coil current and flux through the magnetic circuit, use must be made of the magnetisation curve of the iron. In our simple case we must commence with a chosen value of flux, from which it is possible to estimate B at any point from the local area of cross-section.

For the airgap, $H = B/\mu_o$ and the contribution to a line integral of H around a flux line in the airgap is H x length of path. A similar procedure must be adopted for the iron, but deriving H from the iron characteristic for values of B pertinent to the various lengths of path. The summation of $\sum$ (H x path length) = §**H**.dl = Ni = mmf of coil. Note that, since it is flux which is continuous round the circuit, we can only work from flux to mmf and not vice-versa.

By taking a range of values of flux we can derive a relationship between coil current and flux around the magnetic circuit, which in general will be non-linear.

Fig. 1.3 shows a less crude approximation in that some allowance has been made for the fact that flux can exist in all regions round the coil, particularly from a to a' and b to b'. The flux distribution assumed remains simplified but for some geometries can yield satisfactory results.

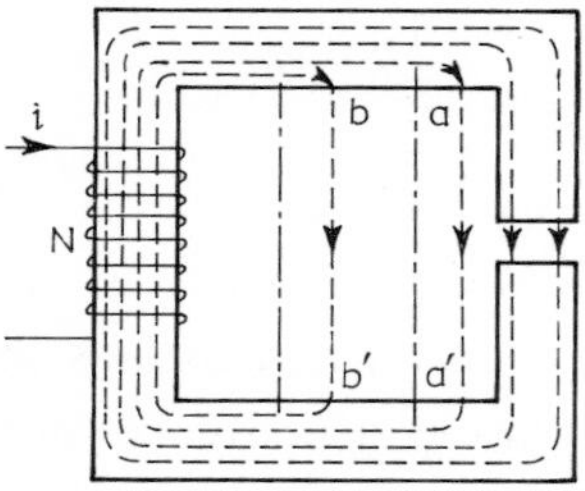

Fig. 1.3.
Improved approximation to flux distribution

Again we must commence with a flux crossing the airgap, and by the same process as previously, we can sum contributions to §**H**.dl until we reach points a and a'. So we can, by assuming the flux distribution shown, calculate H along the line a–a' and hence B in this region. By dividing the region appropriately we can add this flux to that crossing the airgap to obtain a total flux in the core at this point, and a revised (larger) core flux density. From the characteristic for the iron we can determine H, and hence a further contribution to §**H**.dl. The process is repeated for b–b' and its region. The larger the number of regions, the more accurate will be the summation of fluxes and §**H**.dl, provided that the assumed flux distribution is correct.

By even better educated guesses at flux distribution it is possible to arrive at a current-flux relationship which is reasonably close to that which will be measured, and which will be of the form shown by Fig. 1.4. For low values of flux, the iron circuit will have high permeability and the airgap will dominate the magnetic circuit, so that the flux-current relationship initially follows the "airgap line". As the iron saturates, an increasing proportion of the coil mmf is required for the iron, resulting in a "knee" and eventual saturation.

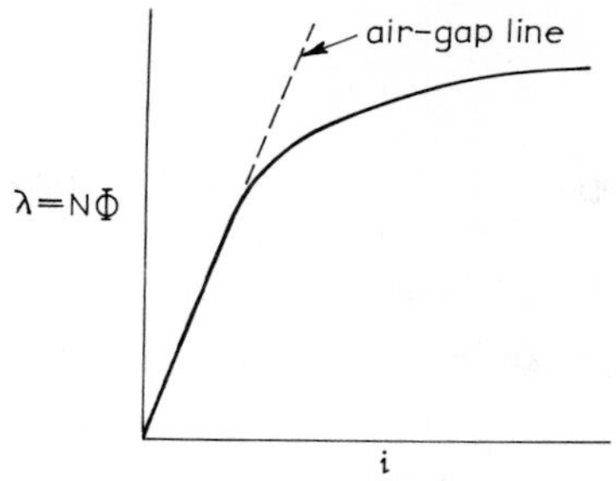

Fig. 1.4. Typical flux linkage –current relationship

The simple example above uses techniques of approximation which are used extensively in machine design. Some account must be taken of fringing (ignored above) when considering flux crossing an airgap between slotted surfaces, but end effects are very often ignored, as above. It is quite common to assume a sinusoidal spatial distribution of flux density around a machine periphery, and to work back from this to determine the mmf required for the iron parts of the circuit.

1.5 Inductance

1.5.1 Self–inductance

Any electric circuit which produces a magnetic field when it carries current has self–inductance. The inductance shows itself as a result of the effects described by Faraday's Law. If the circuit current changes then the magnetic field will change, causing flux linkage with the circuit to change, and inducing a voltage equal to the rate of change of flux linkage.

Consider the simple magnetic circuit of Section 1.4. Let a current i in the coil produce a flux Φ around the magnetic circuit. Each turn of the coil is liked by the flux Φ so, for the complete coil of N turns, the flux linkage is $\lambda = N\Phi$. The inductance is defined as flux linkage per ampere, or

$$L = \lambda/i = N\Phi/i \qquad 1.7$$

Now Φ is a function of the coil mmf Ni, and often the quantity permeance is used in analysis where

$$P = \text{permeance} = \text{flux/mmf} = \Phi/Ni \qquad 1.8$$

or

$$\Phi = PNi \qquad 1.9$$

Hence,

$$L = N^2P \qquad 1.10$$

For changing current,

$$d\Phi/dt = PNdi/dt \qquad 1.11$$

and

$$d\lambda/dt = Nd\Phi/dt$$

$$= N^2Pdi/dt \qquad 1.12$$

From Faraday's Law

$$e = d\lambda/dt = N^2Pdi/dt = Ldi/dt \qquad 1.13$$

When magnetic saturation exists, P is a function of Φ or i. Consequently L is not constant. Then, more correctly,

$$e = d\lambda/dt = (d\lambda/di)(di/dt)$$

and

$$L = d\lambda/di \qquad 1.14$$

A single coil can be represented in circuit diagram form as in Fig. 1.5 , where v = applied voltage

i = coil current

R = resistance

L = inductance

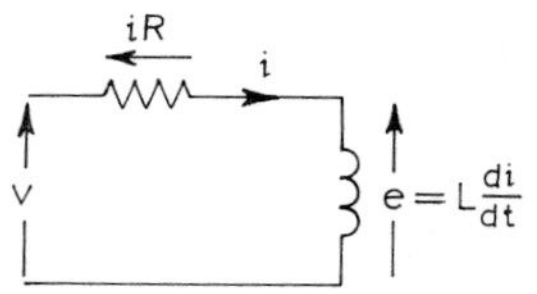

Fig. 1.5.
Circuit diagram with self-inductance

1.5.2 Mutual Inductance

When an electric circuit is linked by flux produced by another circuit, the two circuits are inductively coupled through a mutual inductance.

Mutual inductance between coil 1 and coil 2 is defined as the flux linkage with coil 1 produced per ampere of current in coil 2.

The two circuits may be represented in circuit diagram form as in Fig. 1.6 , where M_{12} is mutual inductance and L_1, L_2 are self-inductances.

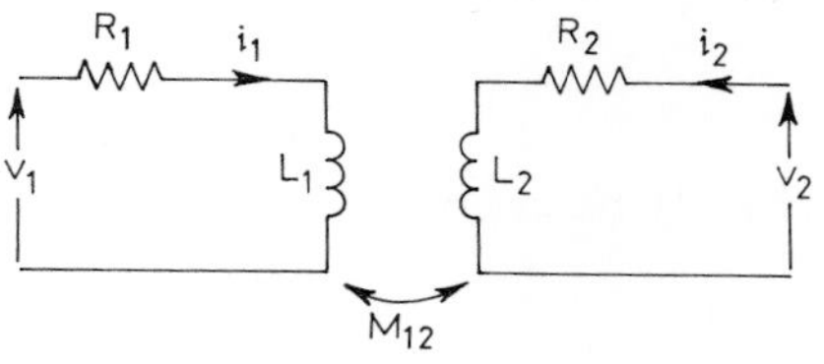

Fig. 1.6. Circuit with mutual inductance

Note that the direction chosen for i_2 is arbitrary. In Fig. 1.6 we regard both coils as power sinks but it is usual in transformers to regard one coil as a power source (i.e. the reverse sense for positive i_2 compared with the above diagram). In rotating machines both alternatives are used, depending upon the machine and the type of winding.

For the above circuit the flux linkages are given by

$$\lambda_1 = i_1L_1 + i_2 M_{12} \qquad 1.15$$

$$\lambda_2 = i_1M_{12} + i_2L_2 \qquad 1.16$$

In a static system the voltage equations are

$$v_1 = i_1R_1 + L_1 \, di_1/dt + M_{12}di_2/dt \qquad 1.17$$

$$v_2 = i_2R_2 + M_{12}di_1/dt + L_2di_2/dt \qquad 1.18$$

1.6 Referred Values of Circuit Parameters and Equivalent Circuits

It is often inconvenient to work in terms of the actual parameters of a coil or circuit. The coil may be inaccessible for measurement and of a complex nature, such as a cage rotor circuit, or it may be of a very different nature from the other coils and working at a different frequency, such as the excitation coil of a synchronous machine. In such cases it is common to work in terms of equivalent parameters referred to a primary coil (or coils), making analysis and parameter measurement more convenient.

Let a primary coil (coil 1) be coupled with another coil (coil 2). For coil 1 we have

$$v_1 = i_1R_1 + d\lambda_1/dt \qquad 1.19$$

where $\lambda_1 = i_1L_1 + i_2M_{12}$

and i_2M_{12} is the component of coil-1 flux linkage produced by current i_2.

As far as effects in coil 1 are concerned, coil 2 could be replaced by another equivalent coil, of any form, with mutual inductance M with coil 1, carrying a current i_2', provided that

$$Mi_2' = M_{12}i_2, \text{ or } i_2' = (M_{12}/M)i_2$$ <u>AT ALL TIMES</u>.

Now i_2 is determined by the voltage equation of coil 2, i.e.

$$v_2 = i_2R_2 + d\lambda_2/dt \qquad 1.20$$

where $\lambda_2 = i_2L_2 + i_1M_{12}$

or

$$v_2 = i_2R_2 + d(i_2L_2)/dt + d(i_1M_{12})/dt \qquad 1.21$$

If the current i_2' in the equivalent coil is to be equivalent to the actual current i_2 at all times, clearly all the parameters which determine i_2 must be modified to equivalent values. Thus

$$v_2' = i_2'R_2' + d(i_2'L_2')/dt + d(i_1M)/dt \qquad 1.22$$

It is convenient to make power invariant. That is, at all times and for all components in the systems, the power in the equivalent system is the same as in the actual system. For this to be so,

$$v_2'i_2' = v_2i_2$$

so that

$$v_2' = (i_2/i_2')v_2 = (M/M_{12})v_2 \qquad 1.23$$

$$(i_2')^2R_2' = (i_2)^2R_2$$

so that

$$R_2 = (i_2/i_2')^2R_2 = (M/M_{12})^2R_2 \qquad 1.24$$

and

$$(i_2')^2L_2' = (i_2)^2L_2$$

so that

$$L_2' = (i_2/i_2')^2L_2 = (M/M_{12})^2L_2 \qquad 1.25$$

With these modifications $i_2'M = i_2M_{12}$ at all times, as required.

The equations can then be written as

$$v_1 = i_1R_1 + L_1di_1/dt + Mdi_2'/dt \qquad 1.26$$

$$v_2' = i_2'R_2' + L_2'di_2'/dt + Mdi_1/dt \qquad 1.27$$

or

$$v_1 = i_1R_1 + (L_1 - M)di_1/dt + Md(i_1 + i_2')/dt \qquad 1.28$$

$$v_2' = i_2'R_2' + (L_2' - M)di2'/dt + Md(i_1+ i_2')/dt \qquad 1.29$$

or

$$v_1 = i_1R_1 + l_1\, di_1/dt + Md(i_1 + i_2')/dt \qquad 1.30$$

$$v_2' = i_2R_2' + l_2'di_1'/dt + Md(i_1 + i_2')/dt \qquad 1.31$$

Here l_1i_1 is a flux linkage associated with i_1 alone and l_2i_2 a flux linkage associated with i_2 alone. These are called "leakage fluxes" with $l_1 = (L_1 - M)$ and $l_2' = (L_2' - M)$ being the leakage inductances.

The pair of equations 1.30 and 1.31 are satisfied by an equivalent circuit of the form shown in Fig. 1.7.

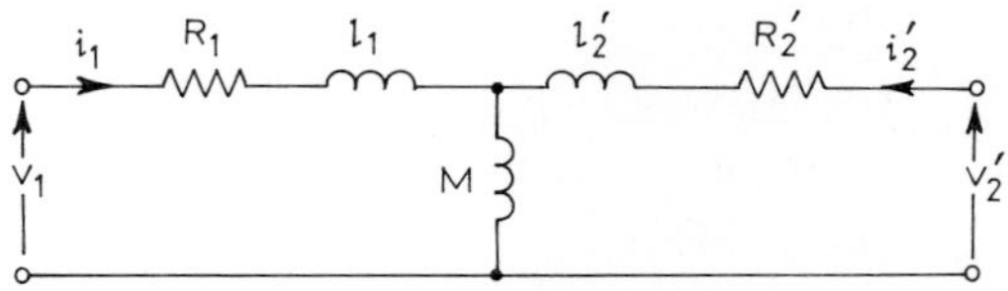

Fig. 1.7. Equivalent circuit of two coupled coils

It is usual to make the equivalent of coil 2 of the same form as the primary coil (coil 1), so that a current i_2' in coil 1 produces the same mmf as a current i_2 in coil 2. Hence, if N_1 and N_2 are the numbers of turns of coils 1 and 2 respectively, then

$$M/M_{12} = N_1/N_2 \qquad 1.32$$

This has the advantage that M represents that part of L_1 produced by flux which also links coil 2. $(L_1 - M) = l_1$ is a leakage inductance associated with flux which does not link coil 2 and can be readily interpreted in terms of leakage flux paths.

If M/M_{12} is chosen to be other than N_1/N_2, then it is possible for a leakage inductance in an equivalent circuit to take a valid negative value. This is unavoidable in some cases, with more than two coils, but is not common in two-coil systems.

With $M/M_{12} = N_1/N_2$, the equivalent parameters of coil 2 are said to be referred to coil 1, and the following equivalent circuit is used:

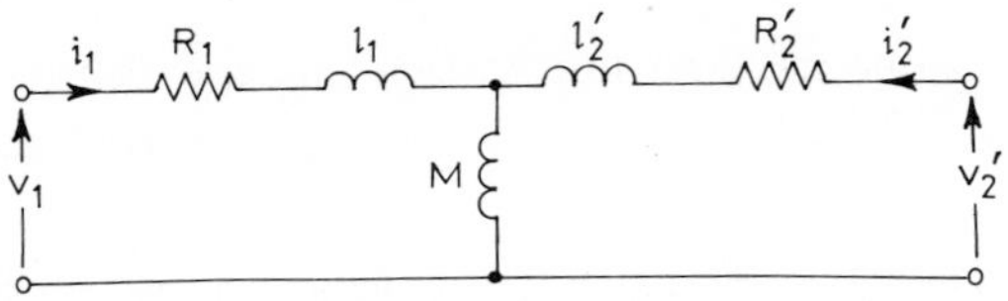

Fig. 1.8. Equivalent circuit referred to coil 1

$i_2' = (N_2/N_1)i_2$

$v_2' = (N_1/N_2)v_2$

$M = (N_1/N_2)M_{12}$

$l_1 = (L_1 - M)$

$l_2' = (N_1/N_2)^2L_2 - M$

$R_2' = (N_1/N_2)^2R_2$

1.7 Generation of EMF

A voltage is being induced in one or more of the coils of any electrical machine if it is producing power. The coil voltage is given by Faraday's Law,

$$e = d\lambda/dt \qquad 1.33$$

where λ is the changing flux linkage of the coil.

λ can change because of a change in:

(a)	current in the coil itself	)
(b)	current in a coupled coil	) transformer effects
(c)	magnetic permeance	)
(d)	relative position between coils	) motional effects

In some machines all four changes occur simultaneously. In all machines producing mechanical power (c) and (d) are essential, and correspond to changes of self- and mutual inductance respectively as functions of position.

Some examples are considered below.

1.7.1 D.C. Machine

The field coil, and any armature (rotor) coil, have a mutual inductance which changes with rotor position. In this case the field is excited with direct current and λ of any armature coil changes at a rate proportional to angular speed, resulting in an alternating voltage in the coil. The alternating coil voltages are rectified to d.c. by mechanical switching of the commutator.

1.7.2 Inductor Machine

In this case the stator is wound with two sets of coils. One set of coils produces a field of long wavelength and the other set is wound with a small wavelength, equal to the pitch of teeth on a rotor. The permeance variation arising from the rotor teeth gives a ripple superimposed on the long-wavelength field and so rotation causes λ of the short-wavelength coils to change, giving an alternating voltage.

1.7.3 Alternator

An alternator has a voltage induced in its a.c. windings by the change in mutual inductance with the rotating field coil, in exactly the same manner as the voltage is generated by a d.c. machine.

When the machines described in Sections 1.7.1 – 1.7.3 are loaded, a further effect occurs associated with the self-inductance of the winding. This effect is sometimes called armature reaction and arises because changing currents produce changing fields and hence inductive voltages.

A three-phase stator winding, with balanced three-phase currents, produces an mmf of virtually constant spatial distribution, travelling at synchronous speed. If the rotor is cylindrical and carries no current, the voltages induced in each phase of the stator are all transformer-type voltages and each phase presents a passive inductive load to a three-phase voltage system, with an effective self-inductance per phase.

If the rotor is salient but without currents, and rotating at synchronous speed, additional voltage components are introduced, associated with the changing self- and mutual inductances. The result is a component of induced voltage per phase which can be "in-phase" with the current. This is the mechanism of the reluctance machine.

1.7.4 Induction Motor

An induction motor, on no load, is in a condition where the self-inductive voltage in any phase is almost equal to the applied voltage, and the flux linkages of each phase combine to produce a travelling flux distribution of virtually constant amplitude.

For a rotor at the same speed as the travelling flux, there will be no change of λ for any rotor circuit. The changing mutual inductances between any rotor circuit and the stator phases are compensated by the phase displacements between the changing stator currents to give a constant rotor flux linkage. However, for any other rotor speed, rotor flux linkage changes occur at slip frequency, causing circulating rotor currents and a rotor mmf, and giving an effect in the stator similar to that experienced in the primary of a transformer when the secondary is loaded.

1.8 Calculation of EMF

Use is sometimes made, in machine design, of the simple equation

$$E = BLv \qquad 1.34$$

to determine the voltage induced in a straight conductor of length L moving with velocity v normal to the plane formed by the direction of flux density B and length L. This can give the correct answer, even for the case where the conductor is deep in a slot, although B is the flux density in the air-gap.

Consider a slotted region with a normal field as shown in Fig. 1.9.

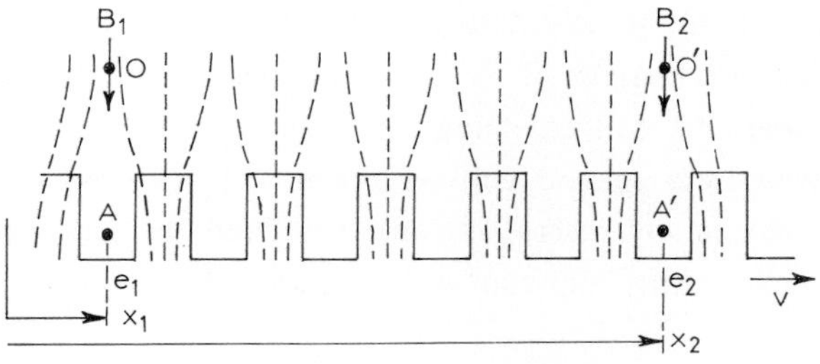

Fig. 1.9. Emf in conductors embedded in slots

It is clear that a coil with sides shown by A,A' is in a region of low flux density, if the teeth are not highly saturated.
The equation

$$\text{coil voltage} = (e_1 - e_2) = (B_1Lv - B_2Lv)$$

gives the voltage for a coil in the airgap with sides shown by O, O'.

The flux linkage of the coil in the slots is given by

$$\lambda = \int_{o}^{x_2} B\,Ldx - \int_{o}^{x_1} B\,Ldx = \int_{x_1}^{x_2} B\,Ldx \qquad 1.35$$

and from Faraday's Law the coil voltage is

$$d\lambda/dt = (d\lambda/dx)(dx/dt) = (d\lambda/dx)v = (B_1 - B_2)Lv \qquad 1.36$$

The equation therefore gives the correct result, provided that the B value is a measure of rate of change of coil flux linkage with position and that B is not changing with time.

Note that a voltage of the type $e = BLv$ does exist but, if used for a coil, the calculation of induced coil voltage must include any effects associated with B changing with time.

It is safest, in general, to calculate a coil voltage from $d\lambda/dt$ where this is the total time derivative of λ and includes both the effects of B changing with time and coil changing position. Suppose the coil to lie in the field produced by a second coil with which it has a mutual inductance M, and let the second coil carry current i_f. Then the flux linkage of the first coil will be $\lambda = Mi_f$, and the voltage induced will be

$$e = d\lambda/dt = d(Mi_f)/dt + i_f(dM/dt) \qquad 1.37$$

The first term represents the effects of B changing with time and the second term represents the effects of the coil changing position, that is transformer emf and motional emf respectively.

1.9 Torque Production

Torque is produced in a rotating machine mainly by the development of magnetic forces acting in a tangential direction on magnetic material. These forces could be calculated by a summation of the magnetic forces on all parts of the rotor or stator, but this would be extremely tedious and requires a detailed solution of the magnetic field in the machine. Account must also be taken of any electromagnetic forces on current-carrying conductors, if such an approach is considered.

Usually a simpler approach is adopted based upon the fact that, at any instant, torque T_e developed by an electromagnetic system is given by

$$T_e = dU/d\theta$$

where U is the energy stored in the magnetic field of the system and θ is an angular displacement.

For a two-coil system with L_1, L_2 and M_{12} such that the energy stored in the field is given by

$$U = (1/2)i_1^2L_1 + (1/2)i_2^2L_2 + i_1i_2M_{12} \quad 1.38$$

then the expression for T_e is

$$T_e = (1/2)i_1^2\partial L_1/\partial\theta + (1/2)i_2^2\partial L_2/\partial\theta + i_1 i_2 \partial M_{12}/\partial\theta \quad 1.39$$

Estimates of inductances, and their variation with angular displacement, can be made to sufficient accuracy by a machine designer by the use of approximate field solutions.

If a conductor of length L, with current i, is in a field B which is normal to the length L, then the conductor experiences a force given by:

$$F = BiL \quad 1.40$$

which is normal to both B and the length L.

This expression is only rarely used by a designer to calculate the torque of an electrical machine, although its use is prevalent in text-books. The expression will only give the correct answer for torque if B is a measure of the rate of change with position of the total flux linkage of a coil, of which the conductor forms a part.

For a conductor in a slot, therefore, the B used must be that in the airgap, normal to the slotted surface, if the force on the slotted surface is required. Use of the true B in the slot will give the actual force on the conductor which is usually, and fortunately, quite small.

The torque expression $T_e = dU/d\theta$, which expresses torque as rate of change of the stored magnetic energy with displacement, is the basis of most torque calculations. It can be shown to be equivalent to calculating torque in the form of the mechanical power developed electromagnetically at a particular speed, where this power is determined as the sum, for all circuits, of $e_g i$ and where i is the circuit current and e_g is the "motional" component of induced circuit voltage generated by movement, or change in self- or mutual inductance.

In the induction motor, with balanced windings and supply voltage, this summated power is equal to 1/s x (ohmic loss in the rotor circuits) where s is per-unit slip.

In the d.c. machine, the summated power is equal to EI_a where I_a is the armature current and E, the generated voltage, is entirely of the type e_g.

For a synchronous machine, the summated power is usually predominantly associated with rotor mmf (excitation mmf) and its consequent effect in producing motional voltage in the stator windings. There may, however, be a further component of motional voltage produced by saliency of the rotor, giving rise to a reluctance torque. Note that, in this type of machine, armature reaction is very important. It can be regarded as causing a winding voltage of the transformer type which does not react with current to produce torque.

CHAPTER 2
Analysis and Design of A.C. Windings

2.1 Polyphase A.C. Windings

The design of a polyphase a.c. winding consists essentially of the definition or description of slot contents, in terms of phase labelling, sense of connection of coil-sides, and number of turns.

A symmetrical m-phase winding comprises m identical phase windings mutually displaced by $2\pi/m$ electrical radians, measured on an angular scale fixed by the wavelength (2 poles) of the fundamental field.

Note that $\theta_{elec} = p\theta_{mech}$
and 2π mech.rad. $\equiv$ $2\pi p$ elec.rad., where p = pole pairs.

The start-points of the three phases of a symmetrical 3-phase winding are mutually displaced by $2\pi/3$ elec.rad. or $2\pi/3p$ mech. rad.

Our treatment will be confined to 3-phase windings but may easily be extended to m phases by analogy. It should, however, be noted that what is commonly called a 2-phase system has phases mutually displaced by 90° in space and time, and hence m = 4.

A single-layer winding is one in which one side of a coil occupies the whole of one slot. A double-layer winding is one in which there are two separate coil-sides in any one slot. In a double-layer winding, a given coil has one coil-side in the upper half of a slot (i.e. the top layer) nearest the air-gap; its other coil-side, or return coil-side, lies in the bottom half of a slot (i.e. in the bottom layer). The angular separation of the two coil-sides forming a given coil is the coil pitch. In many windings, all coils have the same pitch.

A polyphase a.c. winding may be described in terms of several separate sections or "phase belts". Coil-sides in a phase belt carry currents in the same time phase and the angular width of a phase belt in one pole-pitch is the phase spread. In fractional-slot windings, as often used in multipole machines, the average number of slots per phase belt is non-integral. Only integral-slot will be treated here.

For example a 3-phase, double-layer, 120°-spread winding has phase belts which are 120° (elec) wide, as shown in Fig. 2.1., in the developed linear diagram of a 2-pole winding. This distribution would be repeated p times in a winding with p pole-pairs.

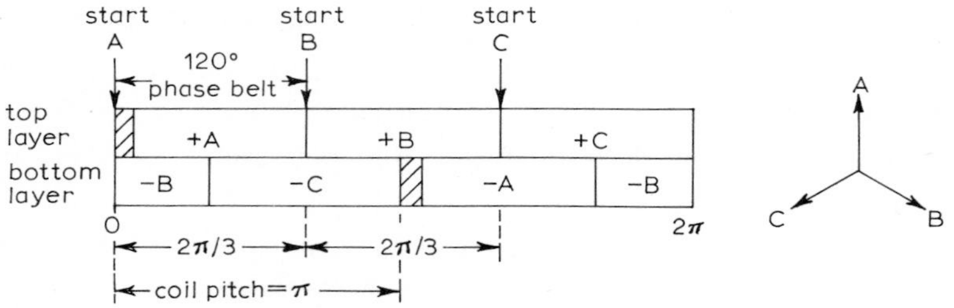

Fig. 2.1. Double-layer winding with 120° phase spread

In this example, the corresponding top (+) and bottom (− or return) coil-sides are separated by π, denoting a coil pitch of π, i.e. a full-pitched winding.

It is noted that the starts of the respective phase winding above are mutually displaced by $2\pi/3$ and occur in the sequence A B C along the top layer. This 120° - spread winding is a true 3-phase winding, but is not commonly used in practice.

A true 3-phase, 120°-spread, single-layer winding is not possible.

The common industrial 3-phase, 60°-spread, double-layer winding is illustrated in Fig. 2.2. This is strictly a 6-phase winding, with the connections to alternate phases reversed [1].

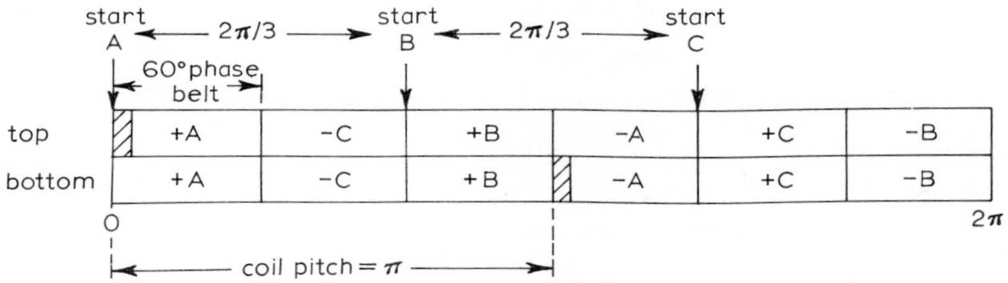

Fig. 2.2. Double-layer winding with 60° phase spread

Note that the phase belts occur in Fig. 2.2 in the sequence A, -C, B, -A, C, -B along the top layer. This is again a full-pitch winding; top and bottom layers in any given slot (i.e. at a given angular position) contain coil-sides in the same phase.

To illustrate a short-pitched coil arrangement, a 2-pole, 3-phase, 60°-spread winding in 12 slots is shown in Fig. 2.3 with a coil pitch of $5\pi/6$ or 150°. The coil pitch may be denoted as slots 1-6 (or 5 slots).

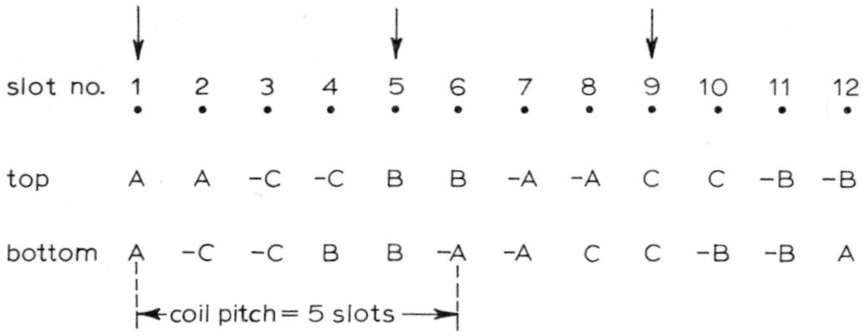

Fig. 2.3. Double-layer winding with $5\pi/6$ coil pitch

In Fig. 2.3,

wavelength = 2 poles = 2π = 12 slot pitches
1 slot pitch = $2\pi/12$ = $\pi/6$ = 30°
phase entries at $2\pi/3$ = 120° = 4 slot pitches
phase spread (or phase-belt width) = 2 slots = 60° = $\pi/3$
coil pitch = 5 slots = 150° = $5\pi/6$

Finally, a single-layer, 3-phase, 60°-spread winding is shown below. This is equivalent to a true 6-phase, 60°-spread winding with alternate phases omitted and the remainder compressed into one layer.

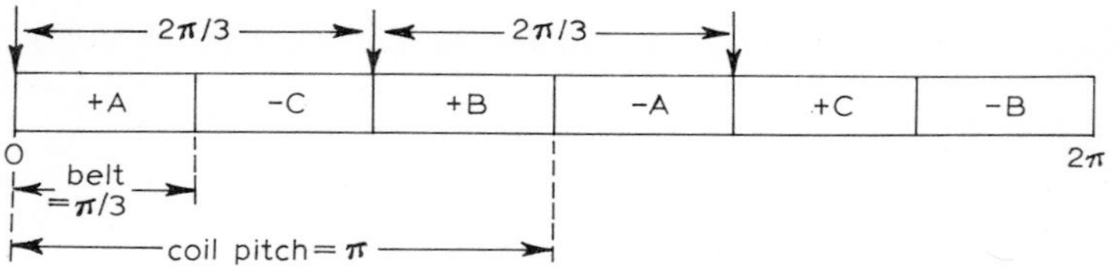

Fig. 2.4. Single-layer winding with 60° phase spread

2.2 Winding Analysis

2.2.1 Generated EMF

The spatial distribution of a synchronously-rotating wave of flux density containing harmonics may be described by

$$B = \sum B_n \sin(n\theta_e + \varphi_n)$$

The r.m.s. induced emf per phase is, for the fundamental component,

$$E_1 = (2\pi/\surd 2)fK_{w1}N\Phi_1 = 4.44nfK_{w1}N\Phi_1 \qquad 2.1$$

and for the nth harmonic

$$E_n = (2\pi/\surd 2)nfK_{wn}N\Phi_n = 4.44nfK_{wn}N\Phi_n \qquad 2.2$$

The fluxes per pole are

$$\Phi_1 = (2/\pi)B_1\tau L \qquad 2.3$$

$$\Phi_n = (2/\pi)B_n(\tau/n)L = \Phi_1(B_n/nB_1) \qquad 2.4$$

Hence

$$E_n/E_1 = B_nK_{wn}/B_1K_{w1} \qquad 2.5$$

Symbols used above are:

θ_e = angle in elec.rad.
N = turns in series per phase
f = fundamental frequency
τ = pole-pitch, at the air-gap
L = core length

K_{wn} = harmonic winding factor (see Section 2.2.3)

$$\text{Total r.m.s.} E = \sqrt{\sum E_n^2}$$

Triplen harmonics produce emfs which are in time phase, so these make no contribution to line voltage and, with a star-connected winding,

$$\text{line } E = \sqrt{3}\sqrt{[\Sigma E_n^2]},\ n \neq 3k \text{ where } k = 1, 2, 3 - - -$$

2.2.2 MMF Analysis

A full-pitched coil carrying an alternating current of r.m.s. value I sets up a pulsating mmf of rectangular waveshape. Since the Fourier analysis of a rectangular wave of amplitude Y is

$$y = (4/\pi)Y[\cos\theta_e + (1/3)\cos 3\theta_e + (1/5)\cos 5\theta_e + - -]$$

the time-maximum amplitude of the rectangular pulsating mmf, in ampere-turns per pole, is

$$F = (4/\pi)(N/2p)\sqrt{2}I[\cos\theta_e + (1/3)\cos 3\theta_e + (1/5)\cos 5\theta_e + - - -]$$

Owing to the symmetry of successive pole-pitches, there are no even harmonics.

Combination of three such pulsating waves, due to three coils symmetrically displaced by 120° (elec.) in space and respectively supplied with balanced 3-phase currents mutually displaced by 120° in time, produces a set of rotating mmf waves defined by

$$F = (3/2)(4/\pi)(N/2p)\sqrt{2}I[\cos(\theta_e - \omega t) + (1/5)\cos(5\theta_e + \omega t) + (1/7)\cos(7\theta_e - \omega t) + (1/11)\cos(11\theta_e + \omega t) + - - - -] \quad 2.6$$

where ω is the fundamental angular frequency of supply.
Note that:

(1) a balanced 3-phase winding produces no triplen harmonics
(2) terms containing $(-\omega t)$ rotate in the same direction as the fundamental wave and those with $(+\omega t)$ rotate in the opposite direction.

Practical windings comprise distributed phase belts, not concentrated in one slot per phase belt, and may also be chorded (i.e. not full-pitched). Winding factors K_{wn} are incorporated to evaluate these aspects, yielding the final expression

$$F = (3/2)(4/\pi)(N/2p)\sqrt{2}I[K_{w1}\cos(\theta_e - \omega t) + (K_{w5}/5)\cos(5\theta_e + \omega t) + (K_{w7}/7)\cos(7\theta_e - wt) + - - - -] \qquad 2.7$$

2.2.3 Winding Factors

It is apparent from the above treatment that the same winding factors K_{wn} are applied in the calculations of both emf and mmf. The validity of this has been formally proved [2].

K_{wn} may be expressed as the product $K_{dn}K_{pn}$ where K_{dn} is the distribution or spread factor, taking account of the distribution of the coil-side in one phase belt, and K_{pn} is the pitch factor, taking account of the coil pitch.

If coil pitch $= 2\alpha$

$$K_{pn} = \sin n\alpha \qquad 2.8$$

For a uniformly distributed (i.e. unslotted) winding with a phase-belt angular width of σ elec.rad.,

$$K_{dn} = \sin(n\sigma/2)/(n\sigma/2) \qquad 2.9$$

For a winding in slots, with q coil-sides per phase belt (i.e. q slots/pole per phase),

$$K_{dn} = \sin(n\sigma/2)/q\sin(n\sigma/q) \qquad 2.10$$

Uniform distribution is equivalent to $q \rightarrow \infty$. Values of K_{d1} are then:

for 60° spread ($\sigma = 60^o$), $K_{d1} = 0.955$
for 120° spread ($\sigma = 120^o$), $K_{d1} = 0.827$.

Values of K_{dn} for 60° -spread, integral-slot windings are tabulated in Table 2.1 for various values of q.

Table 2.1 K_{dn} FOR 60^0-SPREAD INTEGRAL-SLOT WINDINGS

	Number of slots/pole/phase q									
n	2	3	4	5	6	7	8	9	10	∞
1	0.966	0.960	0.958	0.957	0.957	0.957	0.956	0.955	0.955	0.955
3	0.707	0.667	0.654	0.646	0.644	0.642	0.641	0.640	0.639	0.636
5	0.259	0.217	0.205	0.200	0.197	0.195	0.194	0.194	0. 193	0.191
7	0.259	0.177	0.158	0.149	0.145	0.143	0.141	0.140	0.140	0.136
9	0.707	0.333	0.270	0.247	0.236	0.229	0.225	0.222	0.220	0.212
11	0.966	0.177	0.126	0.110	0.102	0.097	0.095	0.093	0.092	0.087
13	0.966	0.217	0.126	0.102	0.092	0.086	0.083	0.081	0.079	0.073
15	0.707	0.667	0.270	0.200	0.172	0.158	0.150	0.145	0.141	0.127
17	0.259	0.960	0.158	0.102	0.084	0.075	0.070	0.066	0.064	0.056
19	0.259	0.960	0.205	0.110	0.084	0.072	0.066	0.062	0.060	0.059
21	0.707	0.667	0.654	0.247	0.172	0.143	0.127	0.118	0.112	0.091
23	0.966	0.217	0.958	0.149	0.092	0.072	0.063	0.057	0.054	0.041
25	0.966	0.177	0.958	0.200	0.102	0.075	0.063	0.056	0.052	0.038
27	0.707	0.333	0.654	0.646	0.236	0.158	0.127	0.111	0.101	0.071
29	0.259	0.177	0.205	0.957	0.145	0.086	0.066	0.056	0.050	0.033
31	0.259	0.217	0.158	0.957	0.197	0.097	0.070	0.057	0.050	0.031
33	0.707	0.667	0.270	0.646	0.644	0.229	0.150	0.118	0.101	0.058
35	0.966	0.960	0.126	0.200	0.957	0.143	0.083	0.062	0.052	0.027
37	0.966	0.960	0.126	0.149	0.957	0.195	0.095	0.066	0.054	0.026
39	0.707	0.667	0.270	0.247	0.644	0.642	0.225	0.145	0.112	0.049
41	0.259	0.217	0.158	0.110	0.197	0.957	0.141	0.081	0.060	0.023
43	0.259	0.177	0.205	0.102	0.145	0.957	0.194	0.093	0.064	0.022
45	0.707	0.333	0.654	0.200	0.236	0.642	0.641	0.222	0.141	0.042
47	0.966	0.177	0.958	0.102	0.102	0.195	0.956	0.140	0.079	0.020
49	0.966	0.217	0.958	0.110	0.092	0.143	0.956	0.194	0.092	0.019
51	0.707	0.667	0.654	0.247	0.174	0.229	0.641	0.640	0.220	0.038
53	0.259	0.960	0.205	0.149	0.084	0.097	0.194	0.955	0.140	0.018
55	0.259	0.960	0.158	0.200	0.082	0.086	0.141	0.955	0.193	0.017
57	0.707	0.667	0.270	0.646	0.172	0.158	0.225	0.640	0.639	0.033
59	0.966	0.217	0.126	0.957	0.092	0.075	0.095	0.194	0.955	0.016
61	0.966	0.177	0.126	0.957	0.102	0.072	0.083	0.140	0.955	0.016
63	0.707	0.333	0.270	0.646	0.236	0.143	0.160	0.222	0.639	0.030
65	0.259	0.177	0.158	0.200	0.145	0.072	0.070	0.093	0.193	0.015

Some particular harmonics merit special attention, because of their magnitude and the importance of the effects which they produce:

(i) Phase-belt harmonics These are associated with the number of phase belts per pole. For the 3-phase winding with three phase belts per pole, these harmonics are of order 5 and 7. In terms of an mmf diagram, these harmonics most nearly fit the approximately triangular differences between the fundamental sinewave and the actual mmf wave.

(ii) First-order slot harmonics These are of order (6q $\pm$ 1) or (2s $\pm$ 1), where s = slots per pole. As seen in Table 2.1, $K_{dn} = K_{d1}$ for these harmonics,. They correspond to the steps in mmf which occur at each slot.

Values of pitch factor K_{pn} are given in Table 2.2, for slots/pole from 9 to 36 (q from 3 to 12), coil pitches from full to two-thirds, and n = 1, 5, 7, 11 and 13.

It is noted that, with two-thirds pitched coils, all K_{pn} = 0.866 and short-pitched coils achieve no reduction of harmonics relative to the fundamental. Certain other coil pitches achieve significant harmonic reduction. Thus, 4/5 pitch eliminates the fifth harmonic, while 6/7 pitch eliminates the seventh harmonic.

The effects of K_{dn} and K_{pn} may be combined to achieve a useful reduction of both fifth and seventh harmonics.

Table 2.2 HARMONIC PITCH FACTORS

	slots pitched	n = 1	n = 5	n = 7	n = 11	n = 13
s = 9	9	1.0000	1.0000	1.0000	1.0000	1.0000
q = 3	8	0.9848	0.6428	0.3420	0.3420	0.6428
	7	0.9397	0.1736	0.7660	0.7660	0.1736
	6	0.8660	0.8660	0.8660	0.8660	0.8660
s = 12	12	1.0000	1.0000	1.0000	1.0000	1.0000
q = 4	11	0.9914	0.7934	0.6088	0.1305	0.1305
	10	0.9659	0.2588	0.2588	0.9659	0.9659
	9	0.9239	0.3827	0.9239	0.3827	0.3827
	8	0.8660	0.8660	0.8660	0.8660	0.8660
s = 15	15	1.0000	1.0000	1.0000	1.0000	1.0000
q = 5	14	0.9945	0.8660	0.7431	0.4067	0.2079
	13	0.9781	0.5000	0.1045	0.6691	0.9136
	12	0.9511	0.0000	0.5878	0.9511	0.5878
	11	0.9135	0.5000	0.9781	0.1045	0.6691
	10	0.8660	0.8660	0.8660	0.8660	0.8660
s = 18	18	1.0000	1.0000	1.0000	1.0000	1.0000
q = 6	17	0.9962	0.9063	0.8192	0.5736	0.4226
	16	0.9848	0.6428	0.3420	0.3420	0.6428
	15	0.9659	0.2588	0.2588	0.9659	0.9659
	14	0.9397	0.1736	0.7660	0.7660	0.1736
	13	0.9063	0.5736	0.9962	0.0872	0.8192
	12	0.8660	0.8660	0.8660	0.8660	0.8660
s = 21	21	1.0000	1.0000	1.0000	1.0000	1.0000
q = 7	20	0.9972	0.9304	0.8660	0.6782	0.5606
	19	0.9888	0.7314	0.5000	0.0802	0.3714
	18	0.9748	0.4305	0.0000	0.7869	0.9770
	17	0.9553	0.0698	0.5000	0.9871	0.7242
	16	0.9304	0.3007	0.8660	0.5519	0.1650
	15	0.9003	0.6293	1.0000	0.2385	0.9092
	14	0.8660	0.8660	0.8660	0.8660	0.8660
s = 24	24	1.0000	1.0000	1.0000	1.0000	1.0000
q = 8	23	0.9978	0.9496	0.8696	0.7518	0.6593
	22	0.9914	0.7934	0.6088	0.1305	0.1305
	21	0.9807	0.5555	0.1950	0.5555	0.5555
	20	0.9659	0.2588	0.2588	0.9659	0.9659
	19	0.9469	0.0654	0.6593	0.8969	0.4423
	18	0.9239	0.3827	0.9239	0.3827	0.3827
	17	0.8969	0.6593	0.9978	0.3215	0.9469
	16	0.8660	0.8660	0.8660	0.8660	0.8660

Table 2.2 (continued)

	slots pitched	n = 1	n = 5	n = 7	n = 11	n = 13
s = 27 q = 9	27	1.0000	1.0000	1.0000	1.0000	1.0000
	26	0.9983	0.9580	0.9182	0.8021	0.7274
	25	0.9933	0.8355	0.6862	0.2868	0.0582
	24	0.9848	0.6428	0.3420	0.3420	0.6428
	23	0.9731	0.3960	0.0582	0.8355	0.9933
	22	0.9580	0.1161	0.4489	0.9983	0.8021
	21	0.9397	0.1736	0.7660	0.6428	0.1736
	20	0.9182	0.4489	0.9580	0.2328	0.5495
	19	0.8937	0.6862	0.9933	0.3960	0.9731
	18	0.8660	0.8660	0.8660	0.8660	0.8660
s = 30 q = 10	30	1.0000	1.0000	1.0000	1.0000	1.0000
	29	0.9986	0.9659	0.9336	0.8387	0.7771
	28	0.9945	0.8660	0.7431	0.4067	0.2079
	27	0.9877	0.7071	0.4540	0.1564	0.4540
	26	0.9781	0.5000	0.1045	0.6991	0.9135
	25	0.9659	0.2588	0.2588	0.9659	0.9659
	24	0.9511	0.0000	0.5878	0.9511	0.5878
	23	0.9336	0.2588	0.8387	0.6293	0.0523
	22	0.9135	0.5000	0.9272	0.1045	0.6691
	21	0.8910	0.7071	0.9877	0.4540	0.9877
	20	0.8660	0.8660	0.8660	0.8660	0.8660
s = 33 q = 11	33	1.0000	1.0000	1.0000	1.0000	1.0000
	32	0.9988	0.9719	0.9450	0.8660	0.8145
	31	0.9954	0.8888	0.7859	0.5000	0.3270
	30	0.9897	0.7559	0.5406	0.0000	0.2814
	29	0.9820	0.5801	0.2351	0.5000	0.7859
	28	0.9719	0.3720	0.0950	0.8660	0.9988
	27	0.9595	0.1424	0.4153	1.0000	0.8415
	26	0.9450	0.0950	0.6900	0.8660	0.3720
	25	0.9285	0.3270	0.8888	0.5000	0.2351
	24	0.9097	0.9647	0.9897	0.0000	0.7559
	23	0.8888	0.7237	0.9820	0.5000	0.9954
	22	0.8660	0.8660	0.8660	0.8660	0.8660

Table 2.2 (continued)

	slots pitched	n = 1	n = 5	n = 7	n = 11	n = 13
s = 36	36	1.0000	1.0000	1.0000	1.0000	1.0000
q = 12	35	0.9990	0.9763	0.9537	0.8870	0.8434
	34	0.9962	0.9063	0.8192	0.5736	0.4226
	33	0.9914	0.7934	0.6088	0.1478	0.1305
	32	0.9848	0.6428	0.3420	0.3420	0.6428
	31	0.9763	0.4617	0.0436	0.7373	0.9537
	30	0.9659	0.2588	0.0872	0.9659	0.9659
	29	0.9537	0.0436	0.5373	0.9763	0.6756
	28	0.9397	0.1736	0.7660	0.7660	0.1736
	27	0.9239	0.3827	0.9239	0.3827	0.3827
	26	0.9063	0.5807	0.9962	0.0872	0.8192
	25	0.8870	0.7373	0.9763	0.5373	0.9990
	24	0.8660	0.8660	0.8660	0.8660	0.8660

2.3 A General Method of Determining Harmonic Winding Factors

The following method [2] may be used to determine the nth-harmonic winding factors K_{wn} of any winding, including single-phase windings and irregular arrangements. It is very well-suited to programming for computation.

The general formula for K_{wn} is

$$K_{wn} = \left[\left\{\sum_{c=1}^{C} N_c \sin(n\alpha_c)\sin(n\beta_c)\right\}^2 + \left\{\sum_{c=1}^{C} N_c \sin(n\alpha_c)\cos(n\beta_c)\right\}^2\right]^{1/2} \Big/ \sum_{c=1}^{C} N_c \qquad 2.11$$

where c = a particular coil of winding

C = number of coils per phase

N_c = number of turns comprising coil c

$2\alpha_c$ = pitch or span of coil c

β_c = angular displacement of the axis of coil c from arbitrary datum (say slot 1)

For a winding in which all coils have the same pitch 2α and the same number of turns, the above form of K_{wn} may be simplified since the pitch factor sin $n\alpha$ may be removed, leaving the distribution factor K_{dn} :

$$K_{dn} = [\{\sum_{c=1}^{C}\sin(n\beta_c)\}^2 + \{\sum_{c=1}^{C}\cos(n\beta_c)\}^2]^{1/2}/C \qquad 2.12$$

The expression may be further simplified for cases in which all phase belts are identical (i.e. excluding fractional-slot and irregular windings). For such cases, let a fundamental wave be defined, representing the main operating field of the machine, and let all angles be measured in electrical radians relative to this wave. Subharmonic fields, having wavelengths greater than the fundamental field, will not be produced by such a winding. The analysis may then be performed for a single phase-belt instead of for the complete winding. C is then replaced by the number of coil-sides per phase belt q, i.e. the number of slots/pole per phase.

Then $$K_{dn} = [\{\sum_{c=1}^{q}\sin(n\beta_c)\}^2 + \{\sum_{c=1}^{q}\cos(n\beta_c)\}^2]^{1/2}/q \qquad 2.13$$

Superficially, it would appear from this equation that K_{dn} depends on the pitch of the coil, since β_c depends on the pitch as well as on the position of the start of each coil. In fact, this is not the case.

Putting γ=slot-pitch angle on fundamental scale and letting the successive coil-sides of a phase belt occupy slots number n_1, n_2, n_3 etc., up to q coil-sides,

$$\beta_1 = \alpha + n_1\gamma$$
$$\beta_2 = \alpha + n_2\gamma \quad \text{etc.} \qquad 2.14$$
$$\beta_c = \alpha + n_c\gamma$$

Manipulation leads to

$$K_{dn} = [\{\sum_{c=1}^{q}\cos(nn_c\gamma)\}^2 + \{\sum_{c=1}^{q}\sin(nn_c\gamma)\}^2]^{1/2}/q \qquad 2.15$$

Thus, K_{dn} depends only upon the distribution of the phase belt, described by γ and the values of n_C for the q slots.

It may be shown that, in the case of a uniform integral-slot winding, this expression is identical to the standard equation (2.10) for distribution factor.

An example of the numbering of slots is given in Section 2.4, on interspersed windings.

The above equation has been used to compute values of K_{dn} for the interspersed windings.

2.4 Interspersed Windings

The harmonic distribution factors of a winding, and hence both the harmonic mmfs which it produces and its generation of harmonic emfs owing to harmonic fluxes, are shown by the foregoing analysis to be determined by the physical distribution of slot positions occupied by the coil-sides in a phase belt. Alteration of this distribution will alter the values of K_{dn}. In interspersed windings [3], this is used to change the values of K_{dn} for low-order harmonics.

The technique consists in interchanging one (or more) end coil-sides of one phase belt (of say A) with the corresponding coil-side (or sides) of the adjacent phase belt, and repeating this cyclic interchange throughout the winding. As a result all phase belts are interleaved.

For example, a section of one layer of a normal 3-phase winding with phases A, B and C and q = 5 is shown below. The prime indicates a coil-side connected in the reverse direction to the unprimed coil-sides.

Coil-side	B'	A	A	A	A	A	C'	C'	C'	C'	C'	B	B	B	B	B
Slot No.	0	1	2	3	4	5	6	7	8	9	10	11	12	13	14	15

With this number of coil-sides per phase belt, the simplest form of interleaving is shown below:

Coil side	A	B'	A	A	A	C'	A	C'	C'	C'	B	C'	B	B	B	A'
Slot No.	0	1	2	3	4	5	6	7	8	9	10	11	12	13	14	15

This may conveniently be described as a 1–3–1 interspersed winding, where the hyphens signify the presence of interleaved coil-sides belonging to the adjacent phase belts.

The effect upon the mmf wave may be regarded as a shaping operation, splitting-up the mmf steps within the phase belt so that the largest, low-order, harmonics are reduced.

Computed values of K_{dn} are listed in Table 2.3 for 1–X–1 interspersed windings for $X = 2$ to $X = 7$, corresponding to $q = 4$ to $q = 9$. the fundamental wave ($n = 1$) is included and values of n are covered up to the first-order slot harmonics.

This Table should be compared with that given previously (Table 2.1) for normal 60° - spread integral-slot windings, noting the changes in fundamental and low-order distribution factors for the same number of coil-sides per phase belt.

Table 2.3. K_{dn} FOR 1-X-1 WINDINGS

n	1-2-1	1-3-1	1-4-1	1-5-1	1-6-1	1-7-1
1	0.8924	0.9149	0.9271	0.9345	0.9393	0.9425
5	0.0990	0.0	0.0563	0.0911	0.1141	0.1300
7	0.2391	0.1182	0.0459	0.0	0.0307	0.0522
11	0.3696	0.2560	0.1713	0.1120	0.0700	0.0397
13	0.3696	0.2890	0.2102	0.1496	0.1048	0.0715
17	0.2391	0.2890	0.2484	0.1983	0.1545	0.1192
19	0.0990	0.2560	0.2484	0.2103	0.1707	0.1364
23	0.8924	0.1182	0.2102	0.2103	0.1868	0.1591
25	0.8924	0.0	0.1713	0.1983	0.1868	0.1648
29		0.9149	0.0459	0.1496	0.1707	0.1648
31		0.9149	0.0563	0.1120	0.1545	0.1591
35			0.9271	0.0	0.1048	0.1364
37			0.9271	0.0911	0.0700	0.1192
41				0.9345	0.0307	0.0715
43				0.9345	0.1141	0.0397
47					0.9393	0.0522
49					0.9393	0.1300
53						0.9425
55						0.9425

Further sets of interspersed windings, which may be useful when the number of coil-sides per phase belt is large, are categorised as 1-1-X-1-1 and 2 --X-- 2. Examples of these, with X = 3, are laid out as follows:

1-1-X-1-1: A B' A B' A A A C' A C' A

2--X--2: A A B' B' A A A C' C' A A

Computed values of K_{dn} for such windings are given in Tables 2.4 and 2.5.

Table 2.4. K_{dn} for 1–1–X–1–1 WINDINGS

n	1-1-3-1-1	1-1-4-1-1	1-1-5-1-1	1-1-6-1-1
1	0.8922	0.9068	0.9169	0.9241
5	0.0619	0.0134	0.0235	0.0518
7	0.1429	0.1039	0.0686	0.0387
11	0.0807	0.1131	0.1158	0.1057
13	0.0229	0.0558	0.0893	0.0988
17	0.2529	0.1185	0.0297	0.0265
19	0.3297	0.2049	0.1008	0.0283
23	0.3297	0.3078	0.2273	0.1444
25	0.2529	0.3078	0.2648	0.1932
29	0.0229	0.2049	0.2648	0.2476
31	0.0807	0.1185	0.2273	0.2476
35	0.1429	0.0558	0.1008	0.1932
37	0.0619	0.1131	0.0279	0.1444
41	0.8922	0.1039	0.0893	0.0283
43	0.8922	0.0134	0.1158	0.0265
47		0.9068	0.0686	0.0988
49		0.9068	0.0235	0.1057
53			0.9169	0.0387
55			0.9169	0.0518
59				0.9241
61				0.9241

Table 2.5. K_{dn} FOR 2 - - X - - 2 WINDINGS

n	2 - - 3 - - 2	2 - - 4 - - 2	2 - - 5 - - 2	2 - - 6 - - 2
1	0.8709	0.8905	0.9040	0.9136
5	0.1663	0.0938	0.0403	0
7	0.2857	0.2145	0.1566	0.1104
11	0.2901	0.2779	0.2485	0.2147
13	0.2132	0.2437	0.2418	0.2246
17	0.0201	0.1058	0.1578	0.1819
19	0.0471	0.0318	0.0978	0.1394
23	0.0471	0.0584	0.0111	0.0424
25	0.0201	0.0584	0.0440	0
29	0.2132	0.0318	0.0440	0.0479
31	0.2901	0.1058	0.0111	0.0479
35	0.2857	0.2437	0.0978	0
37	0.1663	0.2779	0.1578	0.0424
41	0.8709	0.2145	0.2418	0.1394
43	0.8709	0.0938	0.2485	0.1819
47		0.8905	0.1566	0.2246
49		0.8905	0.0403	0.2147
53			0.9040	0.1104
55			0.9040	0
59				0.9136
61				0.9136

2.5 Cage windings

2.5.1 Induced EMF

The r.m.s. induced emf per bar in a cage winding is, from eqn. 1.34,

$$E_b = (1/\surd 2)BLv \qquad 2.16$$

$$v = \omega_2\,(D/2) = (2\pi f_2/p)(D/2)$$

$$\therefore \quad E_b = (2\pi/\surd 2)(B_1LD/2p)\ f_2$$

$$= (4.44B_1LD/2p)sf \qquad 2.17$$

where s = fractional slip and f_2 = sf.

Alternatively, working from the stator emf per phase E_1,

$$\text{rotor emf per turn} = s(E_1/NK_{w1})$$

$$= s(4.44f\Phi_1) \qquad 2.18$$

$$\text{and } E_b = (\text{emf per turn})/2$$

$$\Phi_1 = (2/\pi)B_1\tau L = (2/\pi)B_1L(\pi D/2p)$$

$$\therefore E_b = (4.44B_1LD/2p)sf \text{ as before.} \qquad 2.19$$

2.5.2 Fundamental MMF

Assuming that the rotor slots contain currents which are sinusoidally distributed over a pole-pair, a cage winding may be treated as a polyphase winding. It has to be noted that the conductors under successive pole-pairs are connected in parallel, by the end-rings, rather than in series.

Let S_2 = number of rotor slots
p = pole-pairs
I_b = r.m.s. current per cage bar.

In the same way as in Section 2.2.2 for a 3-phase winding, the fundamental mmf per pole of an m-phase winding is

$$F_1 = (m/2)(4/\pi)(N/2p)K_{w1}\sqrt{2}I_1$$

Note that (N/2p) is the turns per phase acting on one pole.

The cage winding has S_2/p slots per pole-pair, each carrying a current with a different time phase. Three ways of representing this as an m_2-phase winding are:

(i) $m_2 = S_2/2p$, each phase having one full-pitched turn on one pole-pair. The turns/phase per pole = 1/2

(ii) $m_2 = S_2/p$, each phase having 1/2 turn on one pole-pair

(iii) $m_2 = S_2/p$ with one turn per pole, each conductor being half of a cage bar.

In all cases (i) - (iii), K_{d1}, K_{p1} and K_{w1} are all equal to unity.

Adopting representation (i) and substituting $m_2 = S_2/2p$ and $(N/2p) = 1/2$ in the general expression for F_1 yields the fundamental mmf of a cage winding:

$$F_1 = S_2 I_b / \sqrt{2}\pi p \qquad 2.20$$

E_b and I_b, and associated rotor parameters per bar, may be referred to equivalent values per stator phase, as shown in Section 4.3.

2.5.3 End-ring Current and Resistance

Assuming a sinusoidal distribution of bar current over a pole-pitch, the end-ring current also has a sinusoidal space distribution provided the number of bars per pole is greater than, say, 3. The maximum end-ring current $\hat{I}_e$ is then equal to the integral of the bar currents in a half pole-pitch, containing $S_2/4p$ bars,

$$\therefore \quad \hat{I}_e = (2/\pi)(S_2/4p)(\sqrt{2}I_b)$$

and r.m.s. $I_e = S_2 I_b / 2\pi p$ 2.21

If R_e is the resistance of one end-ring, taken around its periphery, the total rotor I^2R loss including both end-rings is:

$$(S_2 I_b^2 R_b + 2 I_e^2 R_e)$$

$$= S_2 I_b^2 R_b + 2(S_2/2\pi p)^2 I_b^2 R_e$$

$$= S_2[R_b + S_2 R_e/2\pi^2 p^2] I_b^2 \qquad 2.22$$

R_b is thus effectively increased by addition of $S_2R_e/2\pi^2p^2$.

Plate 1. Stator of a 3300V, 3-phase, 900kW, 985 rev/min induction motor showing a partly-wound double-layer winding
Courtesy of Laurence Scott & Electromotors Ltd.

CHAPTER 3
MMFs and Excitation Requirements

In order to generate voltages or produce power, a machine requires a magnetic flux around its magnetic circuit linking the electrical circuits or windings. The greater this flux, for a given machine size, the greater the voltage or torque per ampere.

Unfortunately it is not possible to increase machine flux at will. A limit is reached where, because of either large airgaps or more probably magnetic saturation, the necessary magnetising mmf becomes excessive. An excessive mmf is one which requires a winding current so large that either the winding overheats or there is insufficient capability in hand for output power-producing components of current.

3.1 Induction Machines

In this type of machine the airgap is uniform, that is, the rotor is "cylindrical" or "non-salient" and the magnetising mmf is produced by a three-phase stator winding excited by a balanced three-phase voltage system.

Section 2.2 showed that such a winding produces an mmf which is almost a pure sinusoid in its spatial distribution, travelling at constant (synchronous) speed, and with an amplitude of

$$F_1 = (3/2)(4/\pi)(N/2p)K_{w1}\sqrt{2}I = (3\sqrt{2}/\pi)(NK_{w1}/p)I, \text{ amp.turns/pole} \qquad 3.1$$

On no-load, the stator mmf is the only mmf in the machine and is entirely magnetising, that is, producing flux around the magnetic circuit.

Section 2.2 also showed that a three-phase winding of the conventional type only reacts significantly to the fundamental component of the airgap flux distribution. Any distortion of the flux wave gives space harmonics, producing voltage components which are either cancelled by the three-phase connection or greatly attenuated by the winding distribution since usually $K_{wn}/n << K_{w1}$.

The r.m.s. value of the phase voltage is

$$E = (2\pi f/\sqrt{2})NK_{w1}\Phi_1 = 4.44fNK_{w1}\Phi_1 \qquad 3.2$$

where Φ_1 = fundamental flux/pole = $(2/\pi)B_1\tau L$

For a linear machine (i.e. without saturation) and with a smooth airgap of length g, the no-load or magnetising current could be estimated for a given phase voltage by the following process:

$$\text{Flux/pole} = \Phi_1 = E/4.44fNK_{w1} \qquad 3.3$$

$$\text{Peak gap density } B_1 = (\pi/2\tau L)\Phi_1 \qquad 3.4$$

B_1 requires a magnetising field strength B_1/μ_0 in the airgap,

$$\therefore \text{Peak mmf for airgap} = (B_1/\mu_0)g \qquad 3.5$$

Since there is, in this case, no mmf required for the iron, the peak magnetising mmf produced by the stator winding is equal to the peak airgap mmf. Hence, if I_m is magnetising current (rms) per phase,

$$(3\sqrt{2}/\pi)(NK_{w1}/p)I_m = (B_1/\mu_0)g$$

or

$$I_m = (Ep/6fN^2K^2_{w1})(\pi g/\mu_O\tau^2)/L \qquad 3.6$$

In practice, the above equation fails to give an accurate estimate for two main reasons:

(a) the airgap is effectively increased by the presence of slot openings on either or both sides of the airgap

(b) the parts of the magnetic circuit in iron require finite field strength, or a significant mmf component, and magnetic saturation can also distort the flux waveform.

3.1.1 Effective Airgap

Fig. 3.1 shows an approximation to the flux distribution in an airgap between a smooth surface and a slotted surface.

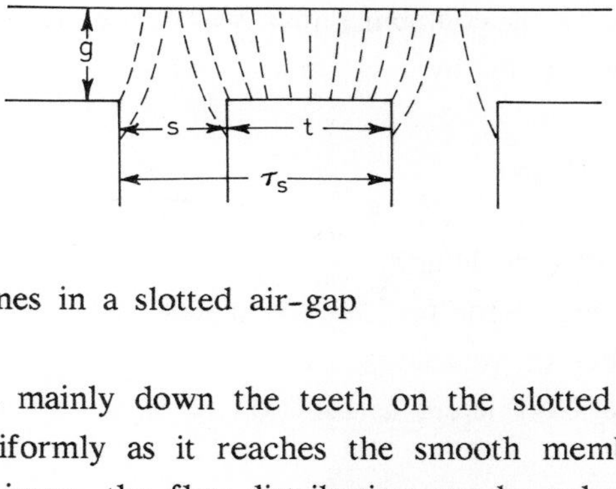

Fig. 3.1. Flux lines in a slotted air-gap

The flux passes mainly down the teeth on the slotted member and then spreads more uniformly as it reaches the smooth member. For a given mmf across the airgap, the flux distribution can be solved analytically, and an average value over a slot pitch τ_s determined.

This problem was originally solved by Carter [4], and a more comprehensive treatment was made recently by Binns [5]. The usual technique is to consider the airgap as having an increased effective length g_e:

$$g_e = k_g g \text{ where } k_g = \tau_s/(\tau_s - ks) \qquad 3.7$$

and k, "the Carter Coefficient", is a function of s/g and s/t.

Most design offices have available curves for Carter Coefficient, which take some account of the differences between "open" slots and "semi-closed" slots.

When both sides of the airgap are slotted, k_{g1} and k_{g2} are determined for both sides and g_e is taken as

$$g_e = k_{g1}k_{g2}g \qquad 3.8$$

or, alternatively, to a better accuracy,

$$g_e = (1/2)\ (k_{g1}k_{g2} + k_{g1} + k_{g2} - 1)g \qquad 3.9$$

The same curves can be used to determine the effective length of the core of the machine if it contains radial ventilating ducts. The effective length of iron is then given by

$$L_i = k_i(L - nw_d k) \qquad 3.10$$

where L = overall core length
k_i = stacking factor for the coreplates
n = number of ventilating ducts of width w_d
k = Carter Coefficient as function of w_d/g

3.1.2 MMF for Core and Teeth

In most designs, the stator and rotor teeth are those parts of the magnetic circuit which experience the highest flux density and hence where magnetic saturation is most pronounced. Before discussing Ampere-turn requirements, we must first deal with the effect of any air paths which might exist in parallel with the iron.

Consider a tooth which has, at a particular point, an iron cross-sectional area A, the flux Φ in a tooth pitch being associated with (tooth + slot + interlaminar air space) of area KA. If the real flux density in the iron is B, then the field strength H will be given by the B-H curve of the iron. However, H also acts on the air in parallel with the iron to give an air flux density of $\mu_0 H$. The flux Φ therefore is given by

$$\Phi = BA + \mu_0 H(K-1)A \qquad 3.11$$

The "apparent" flux density B_{app} is that given by assuming the flux Φ to be carried only by the iron:

$$B_{app} = \Phi/A$$

Hence $B_{app} = B + \mu_0 H(K-1)$

with $K = \text{(iron+air)area/iron area}$ 3.12

Most design offices have curves of the type shown in Fig. 3.2 for various materials and a range of values of K, and these are easily drawn from the true B-H curve.

Such curves are convenient since they save iteration. They become necessary when B_{app} exceeds about 2T.

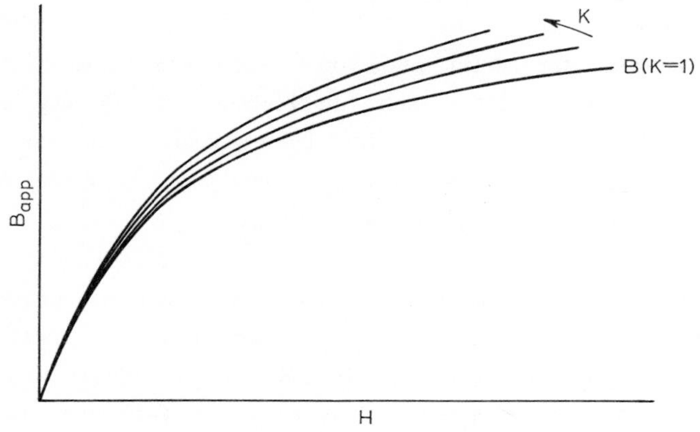

Fig. 3.2. Apparent flux density

Tooth saturation has the additional effect of flattening the top of the flux wave produced by the winding mmf, and so introducing mainly a third harmonic, as shown in Fig. 3.3.

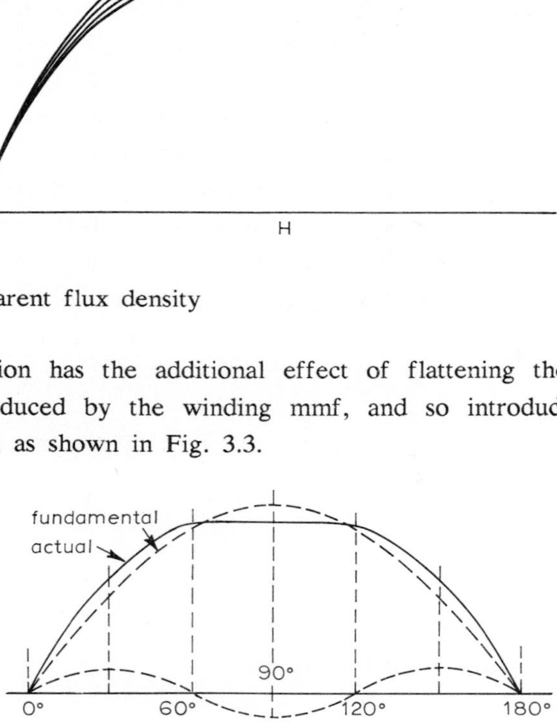

Fig. 3.3. Effect of tooth saturation upon waveshape of air-gap flux density

It is noted from Fig. 3.3 that, at 30° from the peak, the flux density is virtually the same for the fundamental and the actual waveform. A common approximation, which appears to take satisfactory account of distortion, is to calculate the mmf required for airgap and teeth at 30°

from the peak, that is at $\surd 3/2$ of the desired peak fundamental gap flux density B_1, and to multiply this by $2/\surd 3$ to obtain the component of peak fundamental mmf required for airgap and teeth.

In the calculation of magnetising current it is assumed that flux density is predominantly radial in the stator and rotor teeth. This assumption can be justified by analytical solutions, which show peripheral flux to be negligible in the region of the slots compared with that passing radially into the stator or rotor core. Often the teeth of rotor and stator are tapered from tooth tip to slot bottom. This means that, for a given flux per tooth pitch, the apparent density (and the ratio K) varies between tooth tip and slot bottom.

For severe tapers, a process of integration must be used, as indicated in Fig. 3.4 , for a stator tooth. Amplitude of fundamental flux density B_1 is given by

$$B_1 = (\pi/2\tau L)\Phi_1 \qquad 3.13$$

with $$\Phi_1 = E/4.44fNk_{w1} \qquad 3.14$$

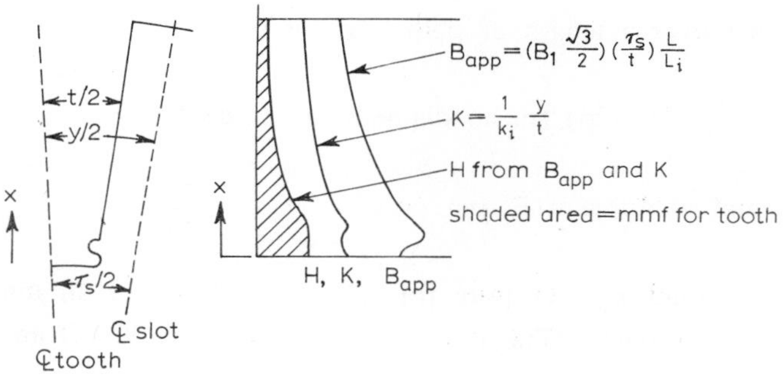

Fig. 3.4. Evaluation of mmf for tapered tooth

A similar process is used for a rotor tooth which will taper the opposite way.

The above method of integration is tedious, and often avoided by an approximation which works well for moderate tapers and saturation levels. This approximation assumes that an average value of H is given at a point one-third of the way up the tooth from the smallest section.

Let the mmf for a stator tooth, for gap density $(\surd 3/2)\ B_1$, be F_{ts} and the mmf for a rotor tooth, for gap density $(\surd 3/2)\ B_1$, be F_{tr}. Then,

$$\text{mmf for gap and teeth} = (g_e/\mu_o)\ (\surd 3/2)\ B_1 + F_{ts} + F_{tr} \qquad 3.15$$

It now remains to calculate the mmf required for the stator and rotor core and this is usually estimated by assuming the peripheral core flux density to be uniform over the depth of core d_c, with a maximum apparent value of

$$B_c = (1/2)\ (\Phi_1/L_i d_c)$$

The factor 1/2 accounts for the fact that the flux/pole divides into two equal halves which travel peripherally in opposite directions around the core.

A value of H is obtained corresponding to this core density, and assumed to act over a length of path l_c given by:

$$l_c = (2/3)(\pi/2p)\ (\text{mean diameter of core})/2 \qquad 3.16$$

to give an mmf component F_c per pole.

This gives an mmf F_{cs} per pole for the stator core and an mmf F_{cr} per pole for the rotor core. The total mmf per pole, at 30° from the peak, is then

$$F_{tot} = (g_e/\mu_o)(\surd 3/2)B_1 + F_{ts} + F_{tr} + F_{cs} + F_{cr} \qquad 3.17$$

Hence, peak fundamental mmf per pole = $(2/\surd 3)F_{tot}$

and the magnetising current is given by:

$$I_m = (\pi/3\surd 2)(p/NK_{w1})\ (2/\surd 3)F_{tot} \qquad 3.18$$

F_{tot} could be written as $(\surd 3/2)F_{gap} + F_{iron}$ 3.19

where F_{gap} = peak fundamental airgap mmf for fundamental flux/pole Φ_1
and $F_{iron} = F_{ts} + F_{tr} + F_{cs} + F_{cr}$ as calculated above.

Now
$$F_{gap} = (g_e/\mu_o)\ B_1 = (g_e/\mu o)(\pi/2)(\Phi_1/\tau L)$$
$$= (g_e\pi/\mu_o\tau 2)(\Phi_1/L) = \Phi_1/\lambda_a L \qquad 3.20$$

where $\lambda_a = (\mu_o\tau/g_e)(2/\pi)$ = airgap permeance per unit length of machine core

and $\Phi_1 = E/4.44fNK_{w1}$

Then, peak fundamental magnetising mmf per pole F_1 can be written as:

$$F_1 = F_{gap}[1 + (2/\surd 3)F_{iron}/F_{gap}] \qquad 3.21$$

and
$$I_m = [Ep/6f(NK_{w1})^2\lambda_a L][1 + (2/\surd 3)F_{iron}/F_{gap}] \qquad 3.22$$

The term inside the second set of square brackets is a factor accounting for the iron part of the circuit and increasing the magnetising current calculated on the basis of the effective airgap permeance only.

The process of calculation may be summarised by the following steps:

(i) calculate g_e and L_i from airgap geometry
(ii) calculate $\lambda_a = \mu_o\tau 2/g_e\pi$
= (fundamental flux/pole per metre length of machine)/(peak fundamental airgap mmf/pole)

(iii) calculate $\Phi_1 = E/4.44fNK_{w1}$

$F_{gap} = \Phi_1/\lambda_a L$

and $B_1 = \Phi_1\pi/2\tau L$

(iv) calculate B_{app} for tooth $= (\sqrt{3}/2)B_1(\tau_s/t)(L/L_i)$

and K for tooth $= (1/k_i)(y/t)$

to determine, either by integration over tooth height or by average, F_{ts} and F_{tr}

(v) calculate B_{app} for core $= (\Phi_1/2)(1/L_i d_c)$

and l_c for core to give F_{cs} and F_{cr}

(vi) determine $F_{iron} = F_{ts} + F_{tr} + F_{cs} + F_{cr}$

(vii) $I_m = [Ep/6f(NK_{w1})^2\lambda_a L]\ [1 + (2/\sqrt{3})\ F_{iron}/F_{gap}]$

The above process yields a value of magnetising current I_m for the phase voltage E induced by flux crossing the airgap. This voltage could be called an "airgap" voltage and will only be equal to the phase terminal voltage V if the stator windings carry no current. In the case of the induction motor, on no-load, we can write the phasor equation

$$\bar{V} = \bar{I}_m r_1 + j\bar{I}_m x_1 + \bar{E} \qquad 3.23$$

where r_1 = resistance per phase

and x_1 = leakage reactance per phase.

Usually $r_1 << x_1$ and $j\bar{I}_m x_1$ is in phase with $\bar{E}$ so that, to a close approximation, $V = E + I_m x_1$.

Strictly, therefore, iteration is required to determine a value for I_m which satisfies the equation $E = V - I_m x_1$ for a given value of V.

However since $I_m x_1$ is typically only about 5% of V, it is often neglected and E is taken as equal to V.

3.1.3 Induction Motor On-Load

When the induction motor is loaded, the rotor windings move with respect to the airgap flux distribution with a relative speed of s times synchronous speed, where s is the fractional slip. Voltages are induced in the rotor at frequency sf_1, causing circulating currents in the rotor windings, which are usually short-circuited, and hence producing a rotor mmf. The rotor mmf, being produced by a polyphase system of circulating rotor currents, will have a predominant component travelling at s times synchronous speed with respect to the rotor and, since the rotor has speed (1−s) synchronous speed, at s + (1−s) = 1 times synchronism speed with respect to the stator. It is thus in synchronism with the airgap flux. The rotor currents will attempt to damp the flux variation causing them, so the rotor mmf will be demagnetising but displaced from the airgap flux by an angle determined by the power-factor of the impedance of the rotor windings. The mmf producing the airgap flux, corresponding to a magnetising current I_m in the stator, is the resultant of the mmfs of stator and rotor windings. The performance of an induction motor is most readily calculated from an equivalent circuit, for which it is necessary to refer rotor quantities to the stator windings.

The basis of referring is that described in Section 1.5, whereby the actual rotor circuits are replaced by static equivalents of the same type as the stator winding, such that the stator winding experiences the same net mmf and flux-linkages. Let the number of rotor phases be m_2, with turns per phase N_2 and fundamental winding factor K_{wr}. An r.m.s. current per rotor phase I_2 will produce a mmf which is almost a sinewave in space, only the fundamental component of which can react with the stator winding, and which will have an amplitude:

$$F_2 = (4/\pi)(m_2/2)(N_2K_{wr}/2p)\sqrt{2}I_2 \text{ , amp.turns/pole}$$

travelling at a speed f_2/f_1 =s times synchronous speed with respect to the rotor windings and at synchronous speed with respect to the stator. Thus, a winding identical to the stator winding would require an r.m.s. current I_2' per phase, at frequency f_1, to produce the same effect, with

$$(4/\pi)(3/2)(NK_{w1}/2p)\sqrt{2}I_2' = (4/\pi)(m_2/2)(N_2K_{wr}/2p)\sqrt{2}I_2$$

so that

$$I_2' = (N_2K_{wr}/NK_{w1})(m_2/3)\ I_2 \qquad 3.24$$

If the fundamental component of resultant airgap flux per pole is Φ_1, the voltage E_2 induced in a rotor phase is given by

$$E_2 = 4.44f_2\ (N_2K_{wr})\Phi_1 \qquad 3.25$$

whereas the voltage E induced in the equivalent stator winding per phase is given by

$$E = 4.44f\ (NK_{w1})\Phi_1 \qquad 3.26$$

so that $$E_2 = (f_2/f_1)(N_2K_{wr}/NK_{w1})E \qquad 3.27$$

The voltage E_2 circulates the rotor phase current through an impedance

$$\bar{Z}_2 = R_2 + j2\pi f_2 l_2 \qquad 3.28$$

where R_2 = (rotor resistance + any external resistance) per phase
l_2 = (rotor leakage inductance + any external inductance) per phase

or, $$\bar{E}_2 = \bar{I}_2\ \bar{Z}_2 \qquad 3.29$$

In the equivalent stator winding, we must use an equivalent impedance $\bar{Z}_2'$ so that:

$$\bar{E} = \bar{I}_2'\bar{Z}_2' \qquad 3.30$$

or

$$\bar{Z}_2' = \bar{E}/\bar{I}_2' = (\bar{E}/\bar{E}_2)\bar{E}_2(\bar{I}_2/\bar{I}_2')/\bar{I}_2 = \bar{E}_2/\bar{I}_2(E/E_2)(I_2/I_2')$$

$$= \bar{Z}_2(E/E_2)(I_2/I_2')$$

$$\text{or } \bar{Z}_2' = (3/m_2)(NK_{w1}/N_2K_{wr})^2\,(f_1/f_2)\,\bar{Z}_2 \qquad 3.31$$

Now $f_2 = sf_1$, so that:

$$\bar{Z}_2' = (3/m_2)(NK_{w1}/N_2K_{wr})^2(R_2/s + j2\pi f_1 l_2) \qquad 3.32$$

This is usually written as:

$$\bar{Z}_2' = R'_2/s + jx'_2$$

with

$$R_2' = (3/m_2)(NK_{w1}/N_2K_{wr})^2 R_2$$

and

$$x_2' = (3/m_2)(NK_{w1}/N_2K_{wr})^2\,2\pi f_1 l_2 \qquad 3.34$$

Thus, the actual rotor circuit can be replaced by an equivalent static 3-phase winding, identical to the stator winding and magnetically coupled with it at the airgap. Each phase of the stator winding then behaves as a transformer, the secondary of which is short-circuited, with effective resistance R_2'/s and leakage reactance x_2'.

The equivalent circuit of the induction motor, referred to the stator per phase, is therefore similar to that for the transformer and is as shown in Fig. 3.5.

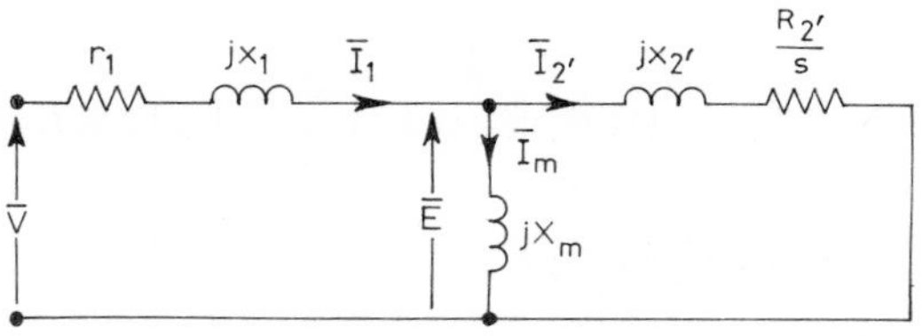

Fig. 3.5. Equivalent circuit of induction motor per stator phase

The airgap flux induces voltage $\bar{E}$ per phase and requires a magnetising current $\bar{I}_m$ per stator phase. The voltage $\bar{E}$ circulates a current, through $\bar{Z}_2'$, which is demagnetising, so that the stator current $\bar{I}_1$ is the phasor sum of $\bar{I}_2'$ and $\bar{I}_m$ (i.e. $\bar{I}_1 = \bar{I}_2' + \bar{I}_m$)

Note that the power loss in the equivalent rotor circuits is 3 $I_2'^2(R'_2/s)$. The actual ohmic loss is $m_2 I_2^2 R_2$ which is equal to 3 $I_2'^2 R'_2$. The apparent discrepancy arises from the frequency transformation. The power $3I_2'^2(R_2'/s)$ is the total power transferred electromagnetically from the stator windings and is comprised of actual ohmic loss in the rotor circuits ($m_2 I_2^2 R_2 = 3I_2'^2 R_2'$) plus power being converted to mechanical form ($3I_2'^2 R_2'(1-s)/s$). This latter corresponds to electromagnetic torque times rotor speed, whereas $3I_2'^2 R_2'/s$ corresponds to electromagnetic torque times synchronous speed.

3.2. Synchronous Machines

Most synchronous machines are of salient-pole construction with four or more poles and with direct-current excitation of the rotor or field winding, and attention will be focussed upon this type. The stator contains a three-phase winding, identical in type to that of an induction motor of the same size. The rotor construction is indicated by Fig. 3.6 for a four-pole machine.

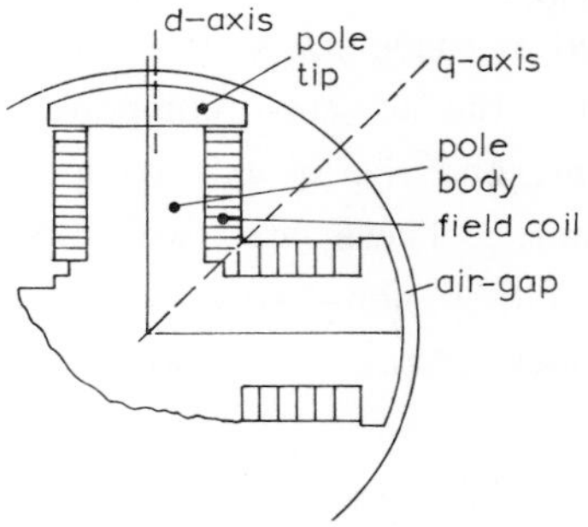

Fig. 3.6. Construction of salient-pole rotor

Sometimes the airgap is graded from g_{min} at pole centre-line to g_{max} at pole-tip edge. Airgaps are much larger than in induction motors.

The magnetic saliency is such that, usually, there are two axes of symmetry, one along the pole centre-line (the direct or d-axis) and the other along the interpolar line (the quadrature or q-axis). Saliency complicates the calculation of performance, because a resultant airgap mmf which is produced by the combined effects of stator and rotor mmfs produces an airgap flux which has a large low-order harmonic content in its spatial distribution and a fundamental component which is not along the axis of the resultant fundamental mmf. It is worth recalling an important property of the three-phase stator winding, namely that it produces virtually a sinewave of mmf and, more importantly, only reacts significantly to the fundamental component of airgap flux distribution. Performance is therefore effected mainly by the fundamental component of airgap flux. Consequently, salient-pole machine analysis is based upon the determination of the contributions to the resultant fundamental airgap flux of mmfs in the d-axis and mmfs in the q-axis. The separation of effects between the two axes is appropriate because usually the two give different flux levels for a given mmf.

Since the rotor winding is centred on the d-axis, only the stator winding can produce an mmf in the q-axis whereas both rotor and stator can produce mmfs in the d-axis. For any load condition, the stator phase current can be considered as made up of two components, with 90° phase difference. These are the d-axis component, giving the d-axis component of stator mmf, and the q-axis component, giving the q-axis component of stator mmf. There are two important operating modes when q-axis mmfs are either non-existent or negligible; these are the cases of stator-winding open-circuit and short-circuit respectively.

3.2.1 Open-circuit Characteristic

The open-circuit characteristic is obtained when only the rotor or field winding is excited. Because of the magnetic configuration, the airgap flux distribution is determined mainly by the airgap geometry. Note that the airgap is usually much larger than in induction motors in order to increase the full-load excitation and hence the ratio of pull-out to full-load torque for constant excitation and stator voltage. A typical synchronous-machine magnetic circuit for the rotor is shown in Fig. 3.8 for a six-pole machine. The distribution of airgap flux density at the stator bore surface will be of the form shown in Fig. 3.7.

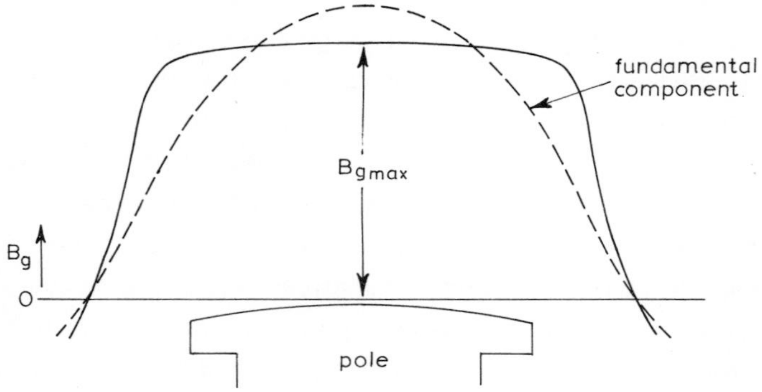

Fig. 3.7. Distribution of air-gap flux density on open circuit

The Wieseman coefficient A_1 is defined as Fundamental amplitude/B_{gmax} and is available [6] in curves of A_1 as a function of gap geometry.

The total flux Φ crossing the airgap per pole will differ from the fundamental flux per pole Φ_1. Wieseman also published curves for K_Φ, defined as Φ/Φ_1.

Hence, for a given fundamental flux per pole, obtained, for a given generated voltage per phase E, from

$$\Phi_1 = E/4.44fNK_{w1} \qquad 3.35$$

the total flux entering the stator per pole is

$$\Phi = K_\Phi \Phi_1 \qquad 3.36$$

and the maximum airgap flux density is

$$B_{gmax} = (\pi/2\tau L)\Phi_1/A_1. \qquad 3.37$$

The airgap mmf can then be calculated as

$$F_{gap} = (g_{mine}/\mu_o)B_{gmax} \qquad 3.38$$

where g_{mine} is an effective airgap at the pole centre-line corrected for stator slot openings. The mmf F_{ts} required for stator teeth is calculated by the method described previously for the induction motor, but using B_{gmax} to calculate apparent tooth flux density:

$$B_{app} = B_{gmax}(\tau_s/t)(L/L_i) \qquad 3.39$$

The mmf F_{cs} required for the stator core is also calculated as for the induction motor, but using Φ rather than Φ_1 to calculate an apparent core flux density:

$$B_c = (1/2)(\Phi/L_1 d_c) \qquad 3.40$$

The sum $(F_{gap} + F_{ts} + F_{cs})$ gives the mmf consumed by airgap and stator core. Further mmf is required for that part of the magnetic circuit on the rotor and this is complicated by interpolar leakage flux.

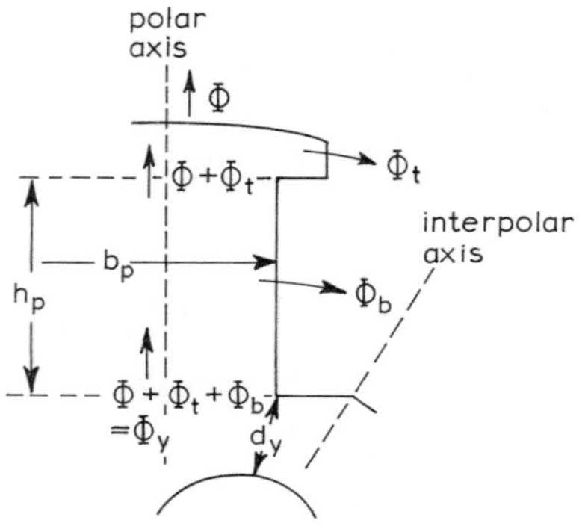

Fig. 3.8. Components of rotor flux

Fig. 3.8 indicates a flux Φ_t which passes from the pole tips of one pole to the tips of the adjacent poles without reaching the stator. This is the pole–tip leakage flux. Assuming the mmf consumed in the pole tip to be negligible, in view of the width available to carry Φ, the mmf per pole driving the leakage flux Φ_t is $(F_{gap} + F_{ts} + F_{cs})$. So

$$\Phi_t = (F_{gap} + F_{ts} + F_{cs})\lambda_{pt} \qquad 3.41$$

where λ_{pt} is a pole–tip leakage permeance, discussed later. The flux Φ_b is a pole-body leakage flux, driven by an mmf per pole which increases with distance from the pole tip owing to the mmf required by the pole body, F_p, and the rotor yoke, F_y. A pole-body leakage permeance λ_{pb}, discussed later, can be estimated from flux plots for the case of infinite permeability for the iron. A suitable approximation is to assume the average mmf acting on λ_{pb} to be given by (F_{gap} + F_{ts} + F_{cs} + (2/3)F_p), so that

$$\Phi_b = (F_{gap} + F_{ts} + F_{cs} + (2/3)\ F_p)\lambda_{pb} \qquad 3.42$$

At this stage we do not know F_p, since this depends upon the pole-body flux density and field strength.

The flux in the pole body is ($\Phi + \Phi_t$) at the tip and ($\Phi + \Phi_t + \Phi_b$) at the root, so that the flux density varies accordingly. Some account can be taken of the highly non-linear nature of the problem by assuming a mean value of flux density over the height of the pole to be given by a flux per pole ($\Phi + \Phi_t$ + (2/3)Φ_b), yielding a mean flux density when divided by the cross-sectional area of the pole body:

$$B_p = \Phi + \Phi_t + (2/3)\Phi_b/Lb_p \qquad 3.43$$

This gives a field strength H_p and an mmf $F_p = H_p h_p$

Thus, iteration is necessary for a value of F_p which gives a suitable value for Φ_b.

The final component of mmf required is F_y for the pole yoke and this is derived from a flux density:

$$B_y = (\Phi + \Phi_t + \Phi_b)/Ld_y \qquad 3.44$$

with a corresponding field strength H_y, giving

$$F_y = H_y l_y \qquad 3.45$$

where l_y = (2/3)(1/2)(π/2p)(mean diameter of yoke)

The total mmf $F_{tot} = (F_{gap} + F_{ts} + F_{cs} + F_p + F_y)$ is equal to $N_f I_f$

where N_f = field turns per pole. This gives a value of I_f for the chosen value of E. Repetition of the process enables E to be plotted as a function of I_f, and this is the Open-Circuit Characteristic (O.C.C.).

The leakage fluxes Φ_t and Φ_b can have a significant effect upon the O.C.C. at higher excitation levels and they are determined by the leakage permeances λ_{pt} and λ_{pb}. Most design offices use expressions for these permeances based upon the work of either Doherty and Shirley [7] or Kilgore [8], with minor empirical modification.

3.2.2 Short-circuit Characteristic

With the stator winding short-circuited at its line terminals, the voltage induced in each phase by the resultant fundamental airgap flux is expended in voltage drops in phase resistance and leakage reactance.

Usually the resistance is small compared with the leakage reactance, so that the power-factor angle of the stator circuit is very close to 90°, with the result that the mmf of the stator is virtually completely demagnetising, that is opposing the field mmf, in the d-axis.

An r.m.s. current I per phase, on short-circuit, thus requires a voltage E equal to Ix_1 generated by airgap flux. This voltage can only be generated by a fundamental flux Φ_1, given by $\Phi_1 = Ix_1/4.44fNK_{w1}$, which will be small compared with that required for rated voltage. Consequently the magnetic circuit will be lightly fluxed and a negligible resultant mmf will be required for the iron. The required field current is

$$I_f = I_{fa} + I_{fx} \qquad 3.46$$

where I_{fa} = field current required to balance stator mmf
and I_{fx} = field current required to generate a voltage Ix_1.

In fact, I_{fa} is the field equivalent of the stator current, such that I_{fa} in the field winding produces the same fundamental airgap flux as a current I in the stator winding on the d-axis.

We have seen that the Wieseman Coefficient A_1 gives the amplitude of the fundamental component of flux density produced by the field winding as

$$B_1 = A_1(\mu_o/g_{mine})N_f I_{fa} \qquad 3.47$$

A similar Wieseman Coefficient A_{d1} gives the amplitude of the fundamental component of flux density produced by a stator mmf acting alone on the d-axis as A_{d1}(peak value), or

$$B_1 = A_{d1}(\mu_o/g_{mine})(3/2)(4/\pi)(NK_{w1}/2p)\sqrt{2}I \qquad 3.48$$

For equivalence of I_{fa} and I, these produce the same fundamental flux or the same value for B_1. Hence

$$I_{fa} = (3\sqrt{2}/\pi)(A_{d1}/A_1)(NK_{w1}/pN_f)\ I \qquad 3.50$$

is the field current equivalent to the armature mmf. If the O.C.C. is available, the component I_{fx} can be obtained from it as that value required to generate a stator phase voltage of magnitude Ix_1.

Alternatively it can be calculated from the mmf $N_f I_{fx}$ required to produce a fundamental airgap flux $\Phi_1 = (Ix_1/4.44fNK_{w1})$, corresponding to a peak gap density equal to $(\pi/2\tau LA_1)\Phi_1$.

Thus

$$I_{fx} = (g_{mine}/2\sqrt{2}fNK_{w1}N_f\mu_o\tau LA_1)Ix_1 \qquad 3.51$$

From the above equations for I_{fa} and I_{fx}, the field current required on short-circuit, $I_f = I_{fa} + I_{fx}$, can be determined for a stator current I. Usually only one point on the Short-Circuit Characteristic (S.C.C.) is calculated because it is nearly always linear, there being negligible mmf required for iron.

Chapter 4 gives an expression for X_{md}, the magnetising reactance of the stator in the d-axis. It can be shown from this that

$$I_{fx}/I_{fa} = x_1/X_{md} \qquad 3.52$$

We could define a d-axis stator current I_e r.m.s. per phase, equivalent to a field current I_f, using the same equivalence criterion as for the reverse process used when deriving I_{fa}. Then

$$I_e = (\pi A_1 p N_f / 3\sqrt{2} A_{d1} N K_{w1})\ I_f \qquad 3.53$$

As far as the stator (d-axis) is concerned, the short-circuit condition may be represented by the equivalent circuit per phase shown in Fig. 3.9.

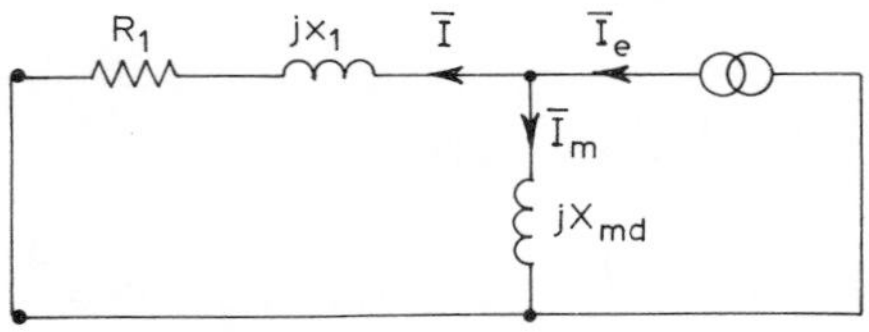

Fig. 3.9. Equivalent circuit on short circuit

$$\bar{I}_e = \bar{I} + \bar{I}_m = \bar{I}(1 + x_1/X_{md}) \text{ if } R_1 \text{ is negligible}$$

so

$$I_f = I_{fa}(1 + x_1/X_{md}) \qquad 3.54$$

on short-circuit.

The current source I_e, representing rotor excitation as seen from the stator, may be replaced by the Thévenin equivalent voltage source, as in Fig. 3.10.

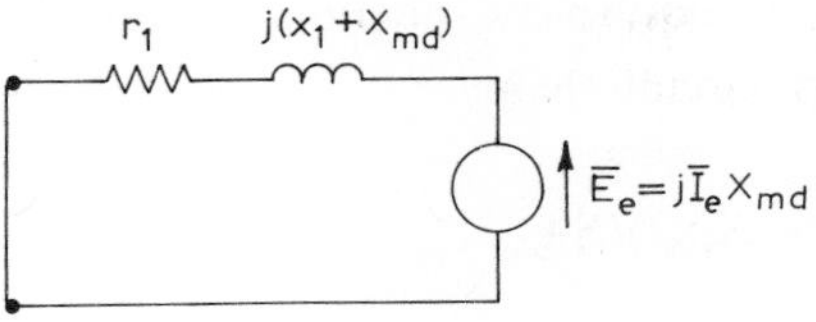

Fig. 3.10. Alternative equivalent circuit on short-circuit

This is a more familiar form and E_e is often called the "excitation voltage".

3.2.3 Short-circuit Ratio

With the O.C.C. and S.C.C. available, the Short Circuit Ratio (SCR) can be readily calculated as

$$SCR = \text{(Field current for rated voltage on open-circuit)}/\text{(Field current for rated current on short-circuit)} \quad 3.55$$

The SCR is an important parameter for synchronous machines, being a measure of the magnitude of the mmf required for magnetisation at rated voltage compared with the magnitude of the mmf of full-load stator current (the armature reaction). For machines on constant-voltage systems the bigger the SCR, the "stiffer" is the machine. That is, the smaller the load angle between resultant flux and d-axis when on load or the smaller the load angle between $\bar{E}_e$ and $\bar{V}$. For generators, a high value of SCR enables the machine to generate at a large leading power-factor angle, if required, without loss of steady-state-stability. For motors it gives a high ratio of pull-out torque to rated torque at constant excitation. The designer increases the SCR by increasing the airgap length.

The SCR is approximately the reciprocal of the per-unit value of saturated, direct-axis synchronous reactance X_{sds}. Thus, in general,

$$X_{sd} = x_1 + X_{md} \qquad 3.56$$

If $I_{es/c}$ is the stator equivalent of field current required for rated stator current I_R on short-circuit then

$$I_{es/c} = I_R \, (x_1 + X_{mdo})/X_{mdo} \qquad 3.57$$

where X_{mdo} is an unsaturated value (i.e. at low flux levels).

If $I_{eo/c}$ is the corresponding stator equivalent of field current for rated volts V_R on open-circuit, then

$$V_R = (I_{eo/c})X_{mds} \qquad 3.58$$

where X_{mds} corresponds to a high (rated voltage) flux condition and thus includes the effects of saturation.

Now $SCR = (I_{eo/c}/I_{es/c})$

so $1/SCR = [x_1 \, (X_{mds}/X_{mdo}) + X_{mds}]/(V_R/I_R)$

$\triangleq x_1 + X_{mds}$, per unit

$= X_{sds}$, per unit

3.2.4 Quadrature-axis Effects

For any load condition the stator current will produce mmf in the q-axis, by its component I_q, as well as in the d-axis, by its component I_d. Alternatively, the existence of a fundamental flux component in the q-axis will require a q-axis component I_q of the stator current since only the stator can produce mmf in the q-axis. The relationship between q-axis flux or the corresponding voltage component and I_q is determined by X_{mq}, the q-axis magnetising reactance. This, in turn, depends predominantly upon the q-axis airgap permeance.

Fig. 3.11 shows the type of flux distribution produced at the stator surface by a q-axis stator mmf.

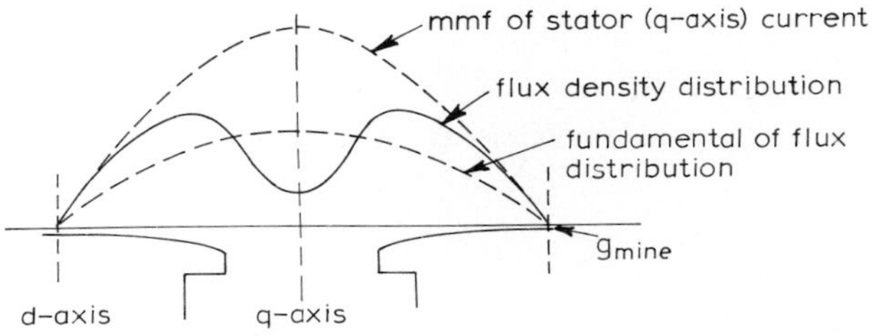

Fig. 3.11. Distributions of q-axis mmf and flux density

The Wieseman Coefficient A_{q1} gives the amplitude of the fundamental component as a fraction of the value which would be given with a uniform airgap of effective length g_{gmine}.

The armature mmf per pole produced by current I_q per phase is

$$F_q = (3/2)(4/\pi)(NK_{w1}/2p)\sqrt{2}I_q \qquad 3.59$$

This gives a fundamental component of flux density

$$B_{q1} = F_q(\mu_o/g_{mine})A_{q1} \qquad 3.60$$

and a fundamental flux per pole

$$\begin{aligned}\Phi_{q1} &= B_{q1}(2/\pi)\tau L \\ &= (2/\pi)\tau L(\mu_O/g_{mine})A_{q1}(3/2)(4/\pi)(NK_{w1}/2p)\sqrt{2}I_q \qquad 3.62\end{aligned}$$

This gives rise to a generated voltage

$$E_q = 4.44fNK_{w1}\Phi_{q1} \qquad 3.63$$

Magnetic saturation is often ignored when considering the q-axis, because the flux is usually much lower than in the d-axis and the large airgap reluctance on this axis dominates the magnetic circuit.

Figs. 3.12 – 3.19 give curves for the evaluation of the Wieseman coefficients as functions of machine design in terms of dimensionless ratios.

Figs 3.12 and 3.13 give coefficients A and B whose product gives the coefficient A_1:

$$A_1 = AB \tag{3.64}$$

Wieseman's A_1 is the same as the C_1 of Kilgore [8].

Figs. 3.14 and 3.15 give coefficients β_1 and β_2 whose product gives the coefficient K_Φ:

$$K_\Phi = \beta_1 \beta_2 \tag{3.65}$$

Figs 3.16 and 3.17 give two further coefficients, also called A and B by Wieseman, whose product gives A_{d1}:

$$A_{d1} = AB \tag{3.66}$$

Wieseman's A_{d1} is the same as the C_{d1} of Kilgore [8].

Figs 3.18 and 3.19 give two more coefficients A and B whose product gives A_{q1}:

$$A_{q1} = AB \tag{3.67}$$

Wieseman's A_{q1} is equivalent to the C_{q1} of Kilgore [8].

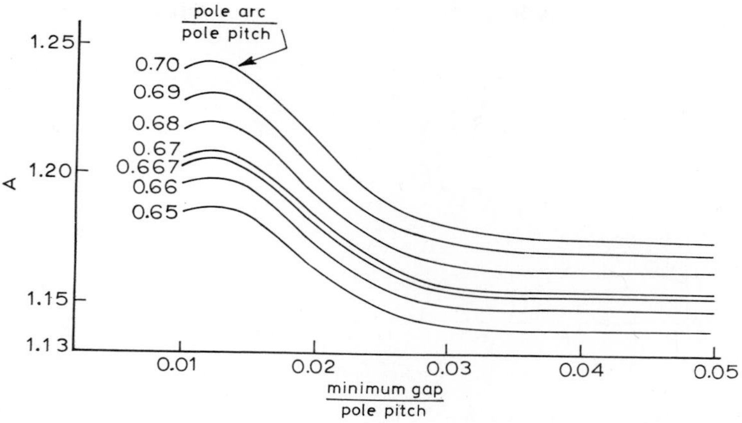

Fig. 3.12. Coefficient A for $A_1 = AB$

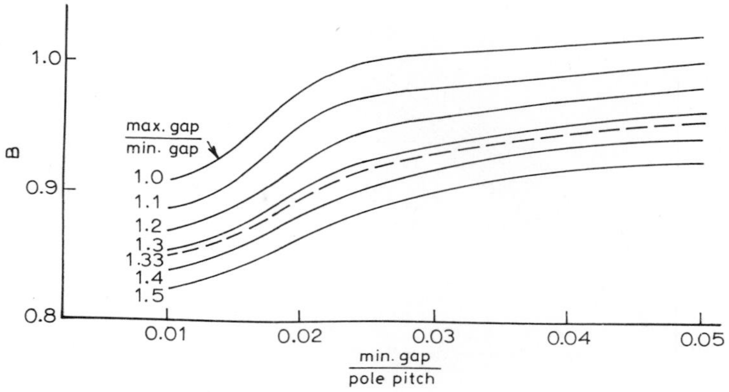

Fig. 3.13. Coefficient B for $A_1 = AB$

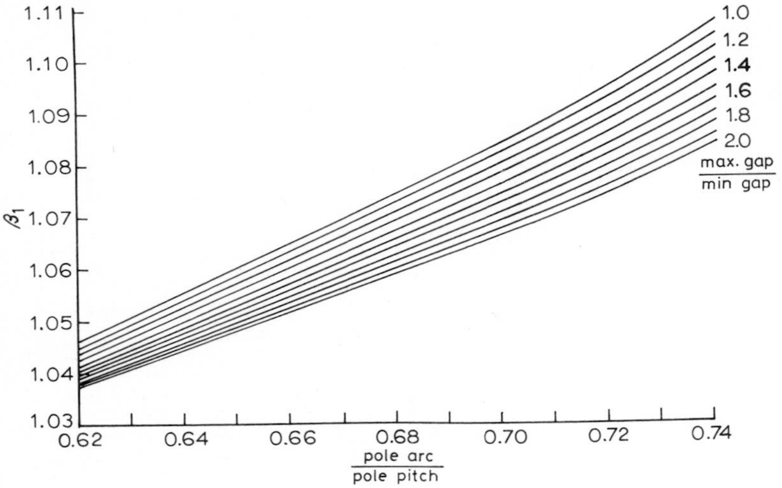

Fig. 3.14. Coefficient β_1 for $K_\Phi = \beta_1 \beta_2$

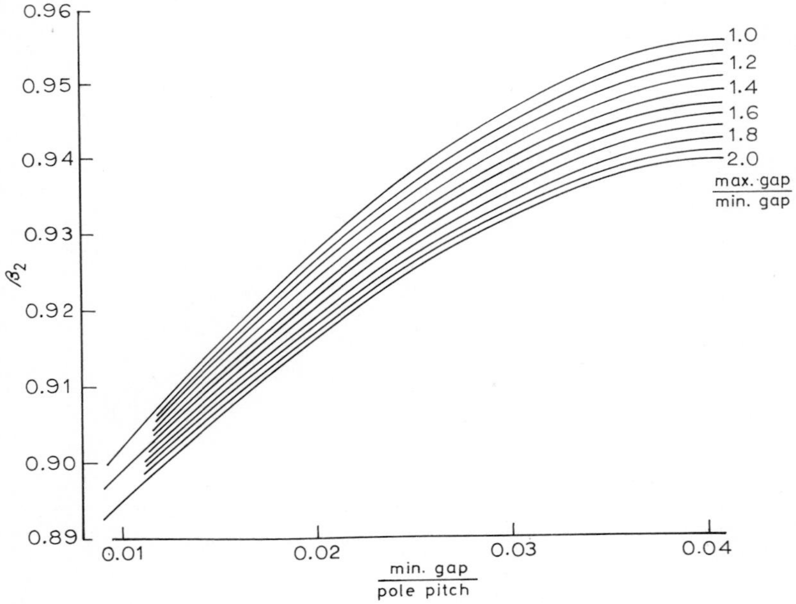

Fig. 3.15. Coefficient β_2 for $K_\Phi = \beta_1 \beta_2$

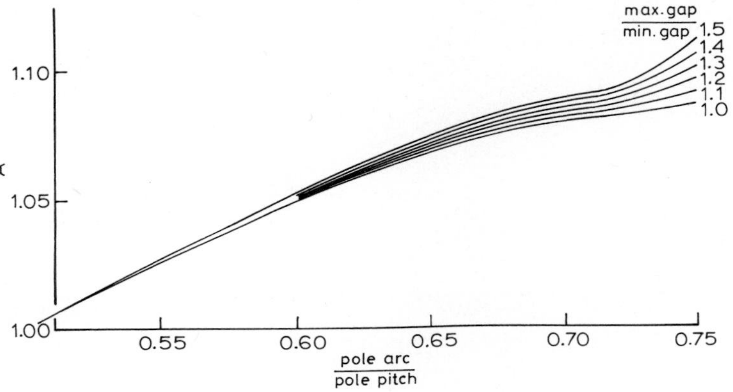

Fig. 3.16. Coefficient A for A_{d1} = AB

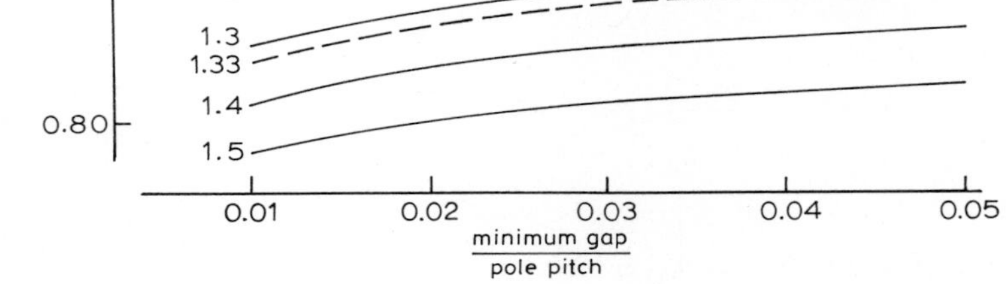

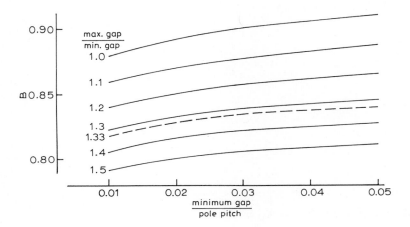

Fig. 3.17. Coefficient B for A_{d1} = AB

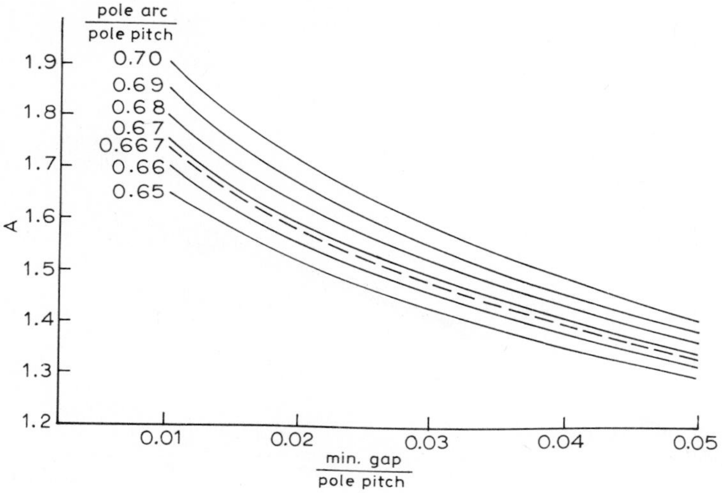

Fig. 3.18. Coefficient A for A_{q1} = AB

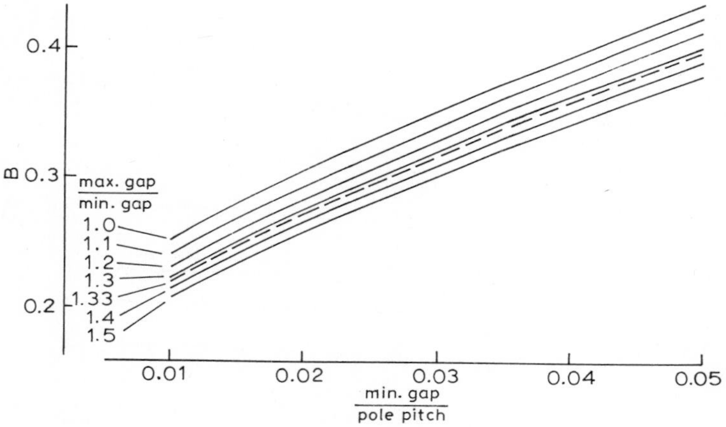

Fig. 3.19. Coefficient B for A_{q1} = AB

Plate 2. Stator of a 3300V, 900kW, 985 rev/min induction motor showing radial ventilation ducts
Courtesy of Laurence Scott and Electromotors Ltd.

Plate 3. A 330MVA, 18kV, 50Hz, 500 rev/min vertical generator/motor for pumped-storage duty
Courtesy of GEC Alsthom Large Machines Ltd.

CHAPTER 4
Permeance and Reactance Calculations

4.1 Magnetising Reactances

4.1.1 Cylindrical-rotor Machine

Considering the magnetisation of an m-phase cylindrical-rotor machine, with magnetising current I_m, we have

$$F_1 = (m/2)(4/\pi)(NK_{w1}/2p)\sqrt{2}I_m \qquad 4.1$$

and $$E = (2\pi/\sqrt{2})NK_{w1}f\Phi \qquad 4.2$$

Hence the magnetising reactance is

$$X_{mc} = E/I_m = [(2m/p)(NK_{w1})^2 f](\Phi/F_1) \qquad 4.3$$

F_1, B and H all have sinusoidal distribution in space:

$$\begin{aligned}\Phi &= (2B/\pi)\tau L \\ &= (2B/\pi)(\pi D/2p)L \\ &= BLD/p \qquad 4.4\end{aligned}$$

With air-gap g,

$$F_1 = Hg = (B/\mu_o)g \qquad 4.5$$

and $\Phi/F_1 = \mu_o DL/pg$

Substituting for Φ/F_1 in eqn. 4.3,

$$X_{mc} = 2m\mu_o(NK_{w1})^2DLf/p^2g \qquad 4.6$$

In practice, an effective airgap g_e is used instead of g, to allow for saturation and slot openings.

4.1.2 Primitive Salient-pole Rotor

It is instructive to develop the following analyses, based on the assumption of purely radial flux in the airgap, and with airgaps of lengths g_1 and g_2 under and between the poles respectively.

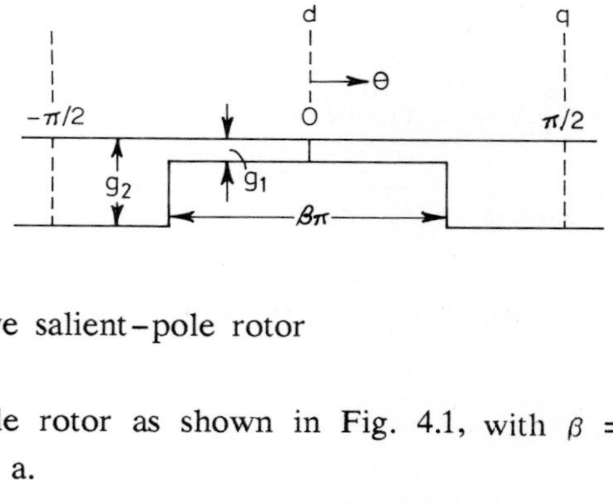

Fig. 4.1. Primitive salient-pole rotor

Consider a 2-pole rotor as shown in Fig. 4.1, with β = pole arc/pole-pitch and g_1/g_2 = a.

Assume that stator currents produce an mmf which is sinusoidally distributed in space and ignore the reluctance of stator and rotor iron.

d-axis magnetising reactance

Let the stator mmf centred on the d-axis be $F\cos\theta$. The distribution of flux density in the airgap is:

$$B(\theta) = \mu_o[F\cos\theta/g(\theta)] \qquad 4.7$$

The fundamental component of $B(\theta)$ is

$$B_{ld} = (4/\pi)\mu_o \int_o^{\pi/2} [F\cos^2\theta/g(\theta)]d\theta \qquad 4.8$$

$$g(\theta) = g_1,\ 0 \leqslant \theta < \beta\pi/2$$

$$g(\theta) = g_2,\ \beta\pi/2 \leqslant \theta \leqslant \pi/2$$

$$B_{ld} = (4\mu_o F/\pi)[\int_o^{\beta\pi/2} \cos^2\theta d\theta/g_1 + \int_{\beta\pi/2}^{\pi/2} (\cos^2\theta)d\theta/g_2]$$

$$= (\mu_o F/\pi)[(\beta\pi + \sin\beta\pi)/g_1 + (\pi - \beta\pi - \sin\beta\pi)/g_2] \qquad 4.9$$

Putting $g_1/g_2 = a$,

$$B_{ld} = (\mu_o F/\pi g_1)[a\pi + (1-a)(\beta\pi + \sin\beta\pi)] \qquad 4.10$$

A cylindrical rotor, with $g_2 = g_1$ and $\beta = 1$, would yield $B_1 = (\mu_o F)/g_1$.

Expressing the d-axis magnetising reactance as

$$X_{md} = k_d X_{mc} \qquad 4.11$$

where X_{mc} is the magnetising reactance of the cylindrical rotor with airgap g_1, the ratio of the flux densities (B_{ld}/B_1) is equal to the d-axis reactance factor k_d.

Thus, $$k_d = (B_{ld}/B_1) = a + (\beta + \sin\beta\pi/\pi)(1-a) \qquad 4.12$$

<u>q-axis magnetising reactance</u>

Apply mmf $F\sin\theta$ centred on the q-axis. Proceeding as before,

$$B(\theta) = \mu_o F\sin\theta/g(\theta) \qquad 4.13$$

and the fundamental component is

$$B_{1q} = (4/\pi)\mu_o \int_o^{\pi/2} [F\sin^2\theta/g(\theta)]d\theta$$

$$= (4/\pi)(\mu_o F)[\int_o^{\beta\pi/2} (\sin^2\theta/g_1)d\theta + \int_{\beta\pi/2}^{\pi/2} (\sin^2\theta/g_2)d\theta]$$

$$= (\mu_o F/\pi)[(\beta\pi - \sin\beta\pi)/g_1 + (\pi - \beta\pi + \sin\beta\pi)/g_2] \qquad 4.14$$

whence $k_q = B_{1q}/B_1 = a + (\beta - \sin\beta\pi/\pi)(1-a)$ 4.15

and $X_{mq} = k_q X_{mc}$ 4.16

The d- and q-axis synchronous reactances are

$$X_d = X_{md} + x_1$$

$$X_q = X_{mq} + x_1$$

where x_1 = stator leakage reactance.

<u>Example</u>. With $\beta = 0.5$, $a = 0.05$, $k_d = 0.83$ and $k_q = 0.22$,
$(k_d/k_q) = 3.77$

Ignoring x_1, with $k_d = 0.83$, I_m of the reluctance motor is 17% greater than that of the cylindrical-rotor induction machine with the same g_1.

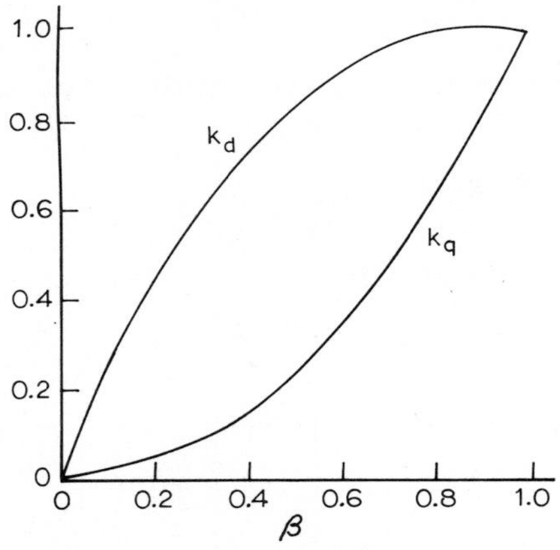

Fig. 4.2. Variation of k_d and k_q with β
$a = 0.034$ or $g_2/g_1 = 30$

Fig. 4.2 shows variations of k_d and k_q with β, for a representative ratio of g_1/g_2.

It may be shown that, ignoring stator resistance, the maximum power-factor of a reluctance motor is given by

$$PF_{max} = [(k_d/k_q) -1]/[(k_d/k_q) +1)] \qquad 4.17$$

For the above example, PF_{max} is 0.58, illustrating the relatively poor power-factor of the primitive reluctance motor.

4.1.3 Rotors with Flux Barriers

High "effective saliency" may be achieved by incorporating flux barriers, rather than by using actual air-gap saliency alone. Some practical rotors for reluctance motors, including flux-barrier types, are shown in Fig. 4.3.

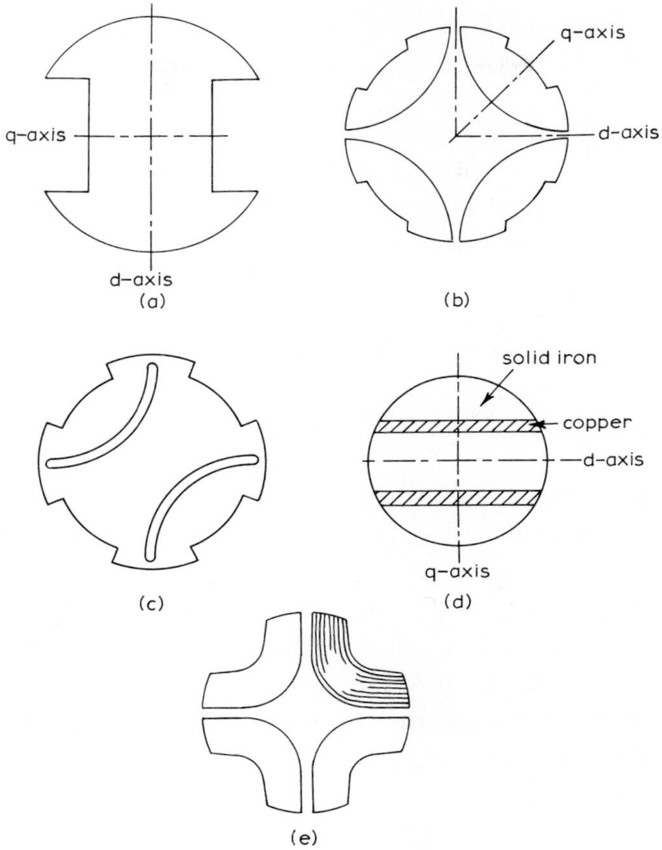

Fig. 4.3. Forms of rotor for reluctance motors
(a) Simple salient-pole type
(b) Segmented rotor with 4 poles [9]
(c) Flux barrier type with 4 poles[10]
(d) Layer-type rotor with 2 poles[11]
(e) Axially-laminated type [12]

The following comments are made, in amplification of Fig. 4.3:

(a) This type derives its saliency from the increased length of airgap on the q-axis. It can be solid or obtained by suitable machining of a cage rotor. X_d/X_q is unlikely to exceed about 4 for this design. Commercial machines often have this construction.

(b) On the d-axis, permeance is little affected by segmentation but on the q-axis the flux reaching the rotor is affected by the intersegment reluctance. It can be solid or laminated with a cage winding. X_d/X_q greater than 7 is possible at realistic flux levels.

(c) Similar to segmented rotor type in characteristics but perhaps easier to construct.

(d) Synchronous characteristics are as those of the barrier or segment types.

(e) Ideal arrangement for high saliency but difficult to construct and to arrange for good asynchronous torque.

The following is an example of an approximate analysis, for a 2-pole machine, demonstrating the mode of operation of flux barriers. It is shown that the reluctance on the q-axis is not simply the added effects of the main airgap and the barrier width.

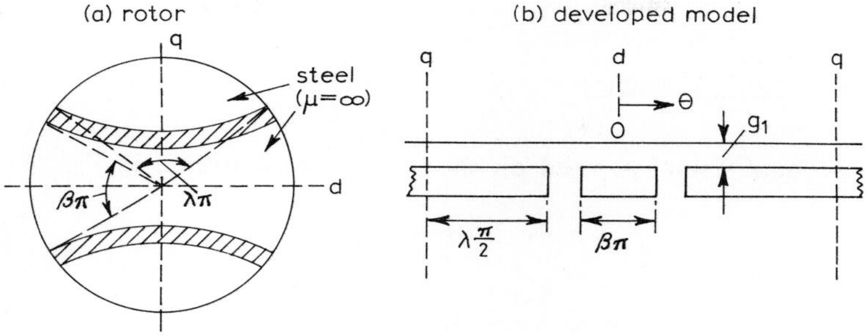

Fig. 4.4. Simple 2–pole rotor with flux barriers

<u>d–axis magnetising reactance</u>

Apply an mmf $F\cos\theta$ centred on the d–axis. Ignore flux entering the rotor at the mouths of the flux barrier. Following the same procedure as for the primitive machine,

$$B_{1d} = (4/\pi)\mu_o\left[\int_o^{\beta\pi/2} F(\cos^2\theta/g_1)\, d\theta + \int_{(1-\lambda)\pi/2}^{\pi/2} F(\cos^2\theta/g_1)\, d\theta\right]$$

$$= (\mu_o F/g_1)[\beta + \lambda + (\sin\beta\pi - \sin\lambda\pi)/\pi] \qquad 4.18$$

Hence $\quad k_d = B_{1d}/B_1 = (\beta + \lambda) + (\sin\beta\pi - \sin\lambda\pi)/\pi \qquad 4.19$

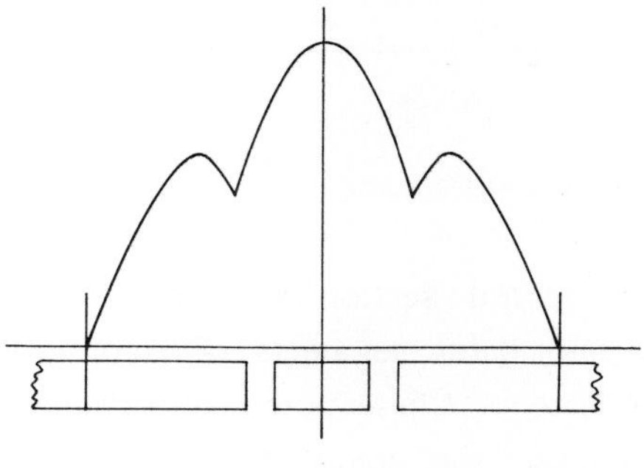

Fig. 4.5. Distribution of flux density with d–axis excitation

The actual flux distribution B_d at the stator surface is as illustrated in Fig. 4.5. There is a local drop in flux density opposite the end of each flux barrier.

q–axis magnetising reactance

Apply the mmf $F\sin\theta$, centred on the q–axis as illustrated in Fig. 4.6:

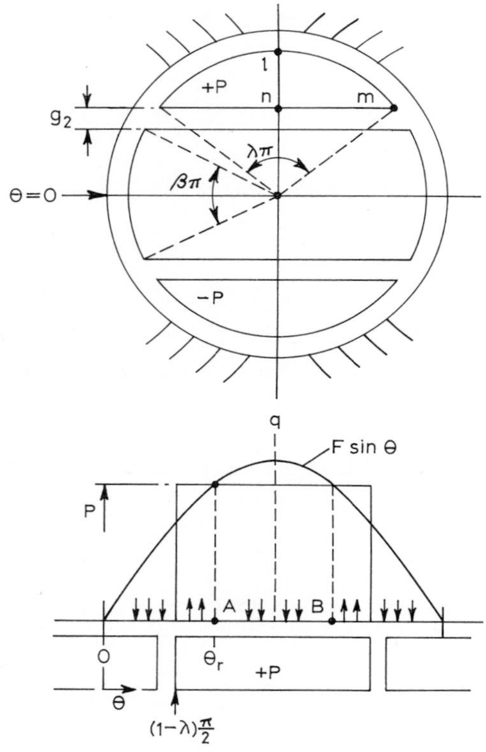

Fig. 4.6. Analysis with q–axis excitation

By symmetry, the central section of rotor iron has zero magnetic potential. The isolated outer segments of rotor iron take up magnetic potentials ± P as indicated. These potentials reduce the flux entering the rotor, since they oppose the stator MMF, and give rise to effective saliency.

Between point A at $\theta=\theta_r$ and point B, $F\sin\theta > P$ and flux passes from stator to rotor in the usual direction. Between $\theta = \theta_r$ and the end of the outer rotor segment at $\theta = (1-\lambda)\pi/2$, $P > F\sin\theta$ and the direction of flux in the airgap is reversed.

To solve for P, the net flux entering the segment through ml is equated to that leaving through mn. Thus, for unit length of machine and rotor radius R,

$$(\mu_o/g_1) \int_{(1-\lambda)\pi/2}^{\pi/2} (F\sin\theta - P)R\,d\theta = P(\mu_o b/g_2) \qquad 4.20$$

where g_2 = barrier width and $b = mn = R\sin(\lambda\pi/2)$

Note g_2, λ and β are related by $g_2 = R[\cos(\lambda\pi/2) - \cos(1-\beta)\pi/2]$

Integrating, eqn. 4.20 yields

$$(FR/g_1)\cos(1-\lambda)\pi/2 - (PR/g_1)\lambda\pi/2 = (Pb/g_2)$$

whence $$P = FR\sin(\lambda\pi/2)/g_1(b/g_2 + R\lambda\pi/2g_1) \qquad 4.21$$

and $$P/F = \sin(\lambda\pi/2)/(bg_1/Rg_2 + \lambda\pi/2) \qquad 4.22$$

The point of flux reversal in the rotor surface lm occurs at $\theta = \theta_r$, where the stator and rotor potentials are equal. That is,

$$P = F\sin\theta_r$$

and

$$\sin\theta_r = P/F \qquad 4.23$$

is given by eqn. 4.22

The flux distribution B_q at the stator surface is then as illustrated in Fig. 4.7.

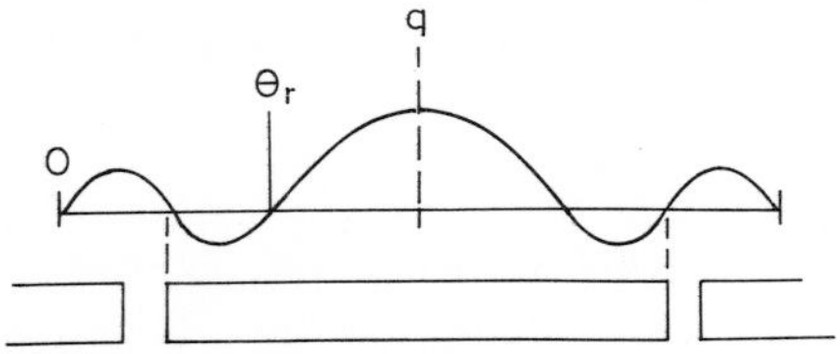

Fig. 4.7. Distribution of flux density with q-axis excitation

B_{1q} and k_q are determined by the usual procedure. Ignoring flux entering the barrier mouth,

$$B_{1q} = (4/\pi)\mu_o[\int_0^{\beta\pi/2} (F\sin^2\theta/g_1)d\theta + \int_{(1-\lambda\pi/2}^{\pi/2} (F\sin\theta - P)(\sin\theta/g_1)d\theta]$$

$$= (\mu_o/\pi g_1)\ [F(\beta\pi - \sin\beta\pi) + F(\lambda\pi + \sin\lambda\pi) - 4P\cos(1-\lambda)\pi/2] \qquad 4.24$$

Dividing B_{1q} by $B_1 = \mu_o F/g_1$ gives

$$k_q = (\beta + \lambda) - (\sin\beta\pi - \sin\lambda\pi)/\pi - (4P/\pi F)\sin(\lambda\pi/2) \qquad 4.25$$

and substituting for P/F,

$$k_q = (\beta + \lambda) - (\sin\beta\pi - \sin\lambda\pi)/\pi - (4/\pi)\sin^2(\lambda\pi/2)/[bg_1/Rg_2 + \lambda\pi/2] \qquad 4.26$$

<u>Example</u> D = 95.2 mm. β = 0.3. g_2 = 2.03 mm. g_1 = 0.41 mm
λ = 0.688. b = 41.4 mm.

$k_d = 0.950$
which is much closer to unity than for primitive rotor
$k_q = 0.208$
$k_d/k_q = 4.57$ and $PF_{max} \triangleq 0.64$

Fig. 4.8 shows variations of k_d and k_q with β for the above example, with various values of g_2.

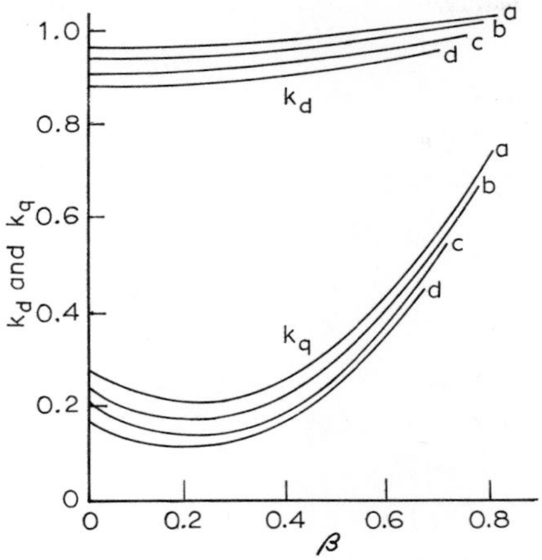

Fig. 4.8. k_d and k_q for a rotor with flux barriers

a: g_2 = 2.03 mm b: g_2 = 3.05 mm

c: g_2 = 4.57 mm d: g_2 = 5.6 mm

Lower k_q and higher k_d/k_q may be achieved by providing channels (i.e. airgap saliency) centred on the q-axis, as indicated in Fig. 4.9.

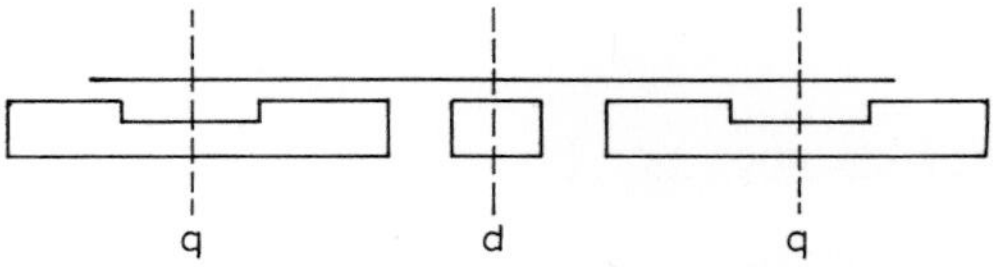

Fig. 4.9. Rotor with both flux barriers and air-gap saliency

Ratios k_d/k_q in the range 5-10 are then obtainable, yielding maximum power-factors up to 0.8

4.2 Leakage-Reactance Calculations

4.2.1 General

From Section 1.5.1 we have permeance $P = \Phi/NI$. With reference to

Fig. 4.10, for an air path of length l and cross-sectional area a, $P = 1/\text{Reluctance} = \mu_o a/l$. Inductance $L = N\Phi/I = N^2P$. For a number of flux paths in parallel, with a common mmf,

$$P = \mu_o \Sigma(a/l)$$

Then $$L = \mu_o N^2 \Sigma(a/l)$$

and reactance $$X = \omega L = 2\pi f L$$

$$= 2\pi f \mu_o N^2 \Sigma(a/l) \qquad 4.27$$

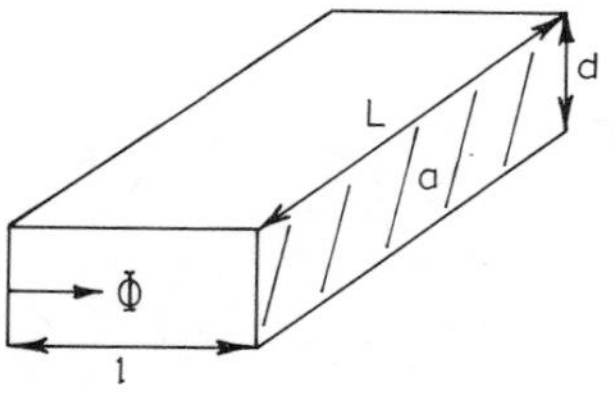

Fig. 4.10. Elementary flux path

In a machine of core length L, specific permeance coefficient (that is, permeance coefficient per unit length) is defined as

$$\lambda = (a/l)/L$$
$$= d/l$$

Then $$X = 2\pi f \mu_o N^2 L \Sigma\lambda \qquad 4.28$$

4.2.2 Leakage Reactance per Phase

Consider firstly a single-layer (or full-pitched double-layer) polyphase winding, with z_s conductors per slot and q slots/pole per phase. There are qz_s turns per pole-pair and the leakage fluxes Φ_l have to cross q slots.

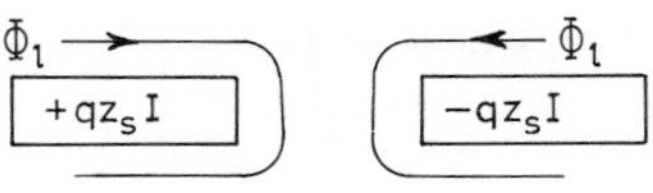

Fig. 4.11. Leakage fluxes per pole-pair

If the specific permeance coefficient per slot is λ_s, the reactance per pole-pair is $2\pi f\mu_o(qz_s)^2 2L(\lambda_S/q)$. If turns of all pole-pairs are connected in series, the leakage reactance per phase is p times the reactance per pole-pair. Or, allowing for parallel connections within the winding,

$$x = 4\pi f\mu_o N^2 L\lambda_s/pq \qquad 4.29$$

where N = turns in series per phase.
Note that the winding actor K_{w1} does not occur here.

This equation is modified subsequently, to cater for additional features.

The following components of leakage reactance may be identified:

1) end-winding leakage of stator and rotor
2) slot leakage of stator and rotor
3) differential leakage (zigzag and belt leakages)
4) skew leakages
5) peripheral air-gap leakage

End-winding leakages x_{e1} and x'_{e2} are dependent upon the particular geometry of the coil-end connections and their proximity to magnetic material. They are usually treated in an empirical manner, for a specific range of designs.

4.2.3 Specific Permeance Coefficient of Slots

Specific permeance coefficients for various shapes of stator and rotor slots have been derived on the assumption that leakage flux passes in straight lines across the slot width. Iron reluctance, skin effect and separation between layers of the winding are usually neglected.

For the slot shape of Fig. 4.12, specific permeance coefficients for the depths d_1, d_2 and d_3 are respectively:

$$\lambda_1 = d_1/w_1 \qquad 4.30$$

$$\lambda_2 = [d_2/(w_2 - w_1)]\log_e(w_2/w_1) \qquad 4.31$$

$$\quad \doteq 2d_2/(w_2 + w_1)$$

$$\lambda_3 = d_3/w_2 \qquad 4.32$$

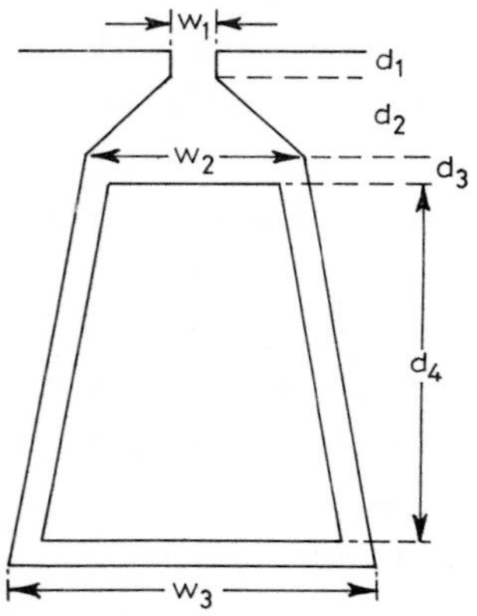

Fig. 4.12. Tapered slot

λ_4 for depth d_4 is found by considering flux in each elemental depth of slot, which links with the current below that element. Summation of the flux linkages, over the depth d_4, yields

$$\lambda_4 = 2d_4/3(w_2 + w_3) \qquad 4.33$$

For a parallel-sided slot, with $w_3 = w_2$,

$$\lambda_4 = d_4/3w_2 \qquad 4.34$$

The above formulae cover both open and semi-closed slots, of rectangular or trapeziodal shape:

$$\text{Total } \lambda_s = \lambda_1 + \lambda_2 + \lambda_3 + \lambda_4 \qquad 4.35$$

For a circular slot λ is independent of slot diameter and, on the same assumption regarding flux direction, is equal to 0.62. This value is commonly increased to 0.66, in view of the actual flux distribution.

For the T-bar slot of Fig. 4.13 , the permeance, evaluated in the same manner as for the trapezoidal slot, is

$$\lambda = [d_5/w_3)(a^2/3) + (d_4/w_2)\ \{a^2 + ab + b^2/3\}]/(a + b)^2 \qquad 4.36$$

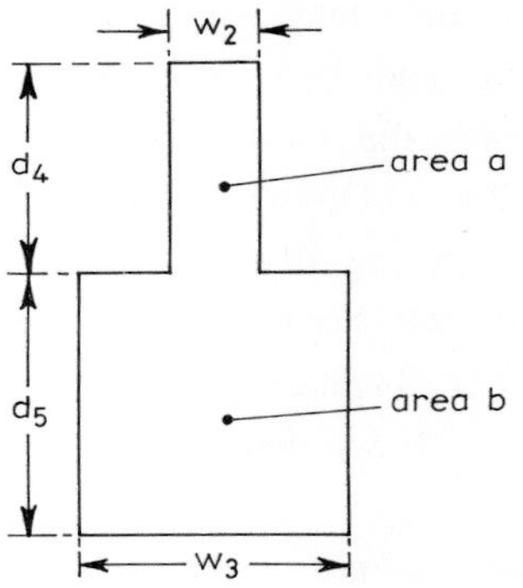

Fig. 4.13. T-snaped slot

T-bar slots are used to exploit skin effect in cage rotors. A treatment of this problem, for general slot shape, is given in Chapter 6.

For windings with short-pitched coils, the above value of λ_s has to be modified [13] to take account of the fact that some slots contain currents having different time phases. For a 60°-spread winding with fractional coil-pitch C_p, λ_s is to be multiplied by k_s where

$$k_s = (1.5\ C_p - 0.25) \text{ for } (1/3) \leqslant C_p \leqslant (2/3)$$
$$\text{and } k_s = (0.75\ C_p + 0.25) \text{ for } (2/3) \leqslant C_p \leqslant 1 \qquad 4.37$$

4.2.4 Differential Leakage Reactance

The differential leakage component of leakage reactance results from the difference between the current distributions on opposite sides of the air-gap. The fundamental components of stator and rotor load mmfs are in balance, but the harmonic components are not. The residual harmonic mmfs produce harmonic fluxes which induce voltages, at the normal frequency, in the originating winding and thus add to the leakage reactance of that winding. The differential leakage reactance x_d of each winding is the ratio of the sum of the voltages induced in that winding, by its differential leakage fluxes, to its fundamental current.

Differential leakage reactance may conveniently be considered as the sum of two parts, the zigzag and belt leakage reactances. Zigzag leakage reactance is associated with the number of slots per pole s and with the slot harmonic orders (2ks ± 1), where k is an integer. Belt leakage reactance is associated with the distribution of a winding in phase belts and thus with the phase-belt harmonics of order (2kb ± 1), where b = phase belts per pole. For 3-phase windings, b = 3 and the phase-belt harmonic orders are (6k ± 1); see Section 2.2.3.

The following observations are are made concerning this distinction between zigzag and belt leakages:

(i) the winding on one side of the air – gap is virtually open-circuited with respect to zigzag leakage fluxes produced by the opposite member in most normal induction motors, which usually have s_1 and s_2 differing by not more than 30%.

(ii) Belt leakage fluxes produced by the stator winding are substantially short-circuited by a cage rotor winding.

(iii) Zigzag leakage is dependent on the number of slots/pole, but not on the coil pitch.

(iv) Belt leakage is independent of slots/pole, but varies with coil pitch.

Precise calculation of zigzag leakage is made difficult by the significant influence of flux fringing around the complicated magnetic geometry of the tooth tops and air-gap. Damping of the zigzag leakage fluxes by induced currents in the winding on the opposite member may usually be neglected, particularly when this member is a skewed cage or is polyphase wound.

The simplest expression for zigzag leakage reactance x_Z of either stator or rotor, with closed slots, is

$$x_Z = \pi^2 X_m / 12 s^2 \qquad 4.38$$

where X_m = fundamental magnetising reactance.

With open stator slots [14], a = total top width/slot pitch and k = actual/effective air-gap length,

$$x_Z = (\pi^2 X_m/12)\ [\{k - a(1 + a)(1 - k)\}/k s_1^2 + 1/s_2^2] \qquad 4.39$$

The separate components x_{Z1} and x_{Z2}, for stator and rotor respectively (referred to a stator phase), are:

$$x_{Z1} = (\pi^2 X_m/12 s_1^2)\ [1 - a(1+a)(1-k)/2k] \qquad 4.40$$

$$x'_{Z2} = (\pi^2 X_m/12)\ [1/s_2^2 - a(1 + a)(1 - k)/2k s_1^2] \qquad 4.41$$

It is noted that, with large s_2, x'_{Z2} may be negative.

Belt leakage reactance x_B of a 3-phase, 60°-spread winding may be evaluated by considering a winding having very large slots per pole s (i.e. a uniformly distributed winding), so that its zigzag leakage reactance is negligible. The main contributing phase–belt harmonics, of orders $n = 5$ and 7, have pole-pitch greater than a slot pitch. The established equation 4.6 for magnetising reactance is therefore used to evaluate x_{Bn} for each harmonic.

Noting that magnetising reactance is inversely proportional to the square of pole number, summation of the belt harmonic contributions yields

$$x_B = (X_m/K_{w1}^2)\ \Sigma K_{wn}^2/n^2,\ n \neq 1$$

$$= \{X_m/(K_{d1}K_{p1})^2\}\ \Sigma(K_{dn}K_{pn}/n)^2 \qquad 4.42$$

Now $K_{dn} = \sin(n\pi/6)/q\sin(n\pi/6q)$

$$= \sin(n\pi/6)/(n\pi/6)[\sin(n\pi/2s_1)/n\pi/2s_1) \qquad 4.43$$

since $s_1 = 3q_1$.

So, for large s_1,

$$K_{dn} = \sin(n\pi/6)/(n\pi/6) \qquad 4.44$$

$$\text{and } K_{dn}/K_{d1} = (1/n)\ \sin(n\pi/6)/\sin\pi/6 \qquad 4.45$$

Hence

$$|K_{dn}/K_{d1}| = 1/n \text{ for } n = (6k \pm 1)$$

$$\therefore\ x_B = (X_m/K_{p1}^2)\ \Sigma(K_{pn}^2/n^4) \qquad 4.46$$

For a full–pitch winding, $x_B = 0.00214\ X_m$.

With a coil pitch of 80%, the fifth harmonic is eliminated and x_B is reduced to $0.00024\ X_m$.

For wound-rotor machines, x_B may be evaluated in this way for both stator and rotor. A cage rotor winding has no phase belts and its belt leakage is zero.

Stator belt-leakage fluxes induce voltages and circulating currents in a cage rotor and, because of their good coupling and high slip, the fluxes are almost totally damped out and x_B is then negligible. The induced harmonic currents, losses and torques may be significant, contributing to stray load losses and loss torque.

4.2.5 Skew Leakage

Rotor slots (or, alternatively, sometimes stator slots) of cage motors are often skewed by one stator slot pitch. A principal objective of this feature is to minimise induced voltage per bar and circulating currents due to both stator slot-harmonic mmfs and pulsations of main flux owing to stator slot openings. There are, however, some disadvantageous effects associated with skewing, including a decoupling effect which may be regarded as equivalent to introduction of a skew-leakage flux.

Referring to Fig. 4.14, the effect of skewing is seen to be an angular displacement of the mmf of the skewed member. The phase difference between the fundamental load mmfs of stator and rotor increases with distance y from the core centre. The resulting difference in mmf is the skew mmf.

The circumferential distribution of stator load mmf is

$$F_1 = F\sin 2px/D \qquad 4.47$$

The rotor mmf F_2 is assumed to balance F_1 at $y = 0$, so its distribution as a function of x and y is

$$F_2 = F\sin\{(2p/D)(x - x_{SK}y/L)\} \qquad 4.48$$

where $x_{SK} = k\pi D/S_1$ and k is skew expressed as a multiple of one stator slot pitch.

$$\text{Skew mmf } F_{SK} = F_1 - F_2$$

$$= 2F\sin\{(p/Dx_{SK})(y/L)\}\cos\{(p/D)(2x - x_{SK}y/L)\} \qquad 4.49$$

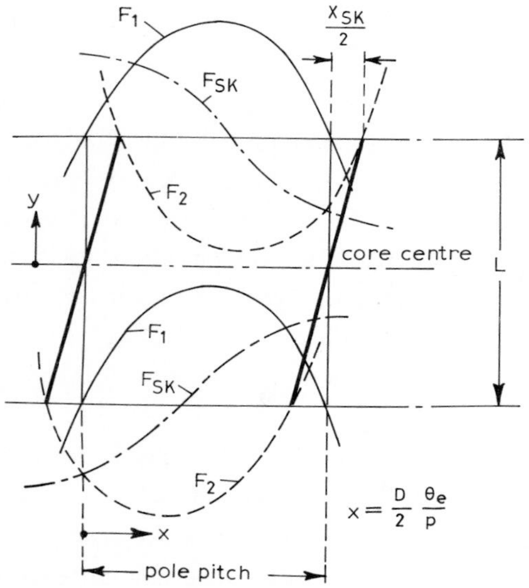

Fig. 4.14. Distribution of MMFs.
MMF waves are shown at the core ends

At distance y from the core centre, the peak skew mmf is

$$F_{SK} = 2F\sin\{(p/D)(x_{SK}y/L)\}$$
$$= 2F\sin\{(\pi pk/S_1)(y/L)\} \qquad 4.50$$

The greatest F_{SK}, at $y = L/2$, is equal to $2F\sin(\pi pk/S_1)$ or $2F\sin(\pi k/4s_1)$.

The coupling may be expressed in terms of the ratio of emf per bar with skew to that without skew. The phasor emf in an elemental length dy at y is $edy\angle(\alpha y/L)$, where e = emf per unit length.

The summation of emfs is shown in Fig. 4.15 and the emf ratio K_{SK} is chord AB/arc AB. Thus,

$$K_{SK} = \sin(\alpha/2)/(\alpha/2)$$

$$= \sin(\pi k/2s_1)/(\pi k/2s_1) \qquad 4.51$$

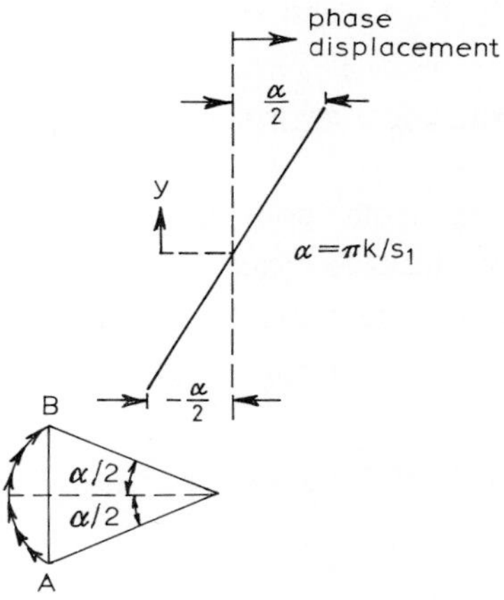

Fig. 4.15. Fundamental skew-leakage emfs

When k = 1, K_{SK} is close to unity for normal values of s_1; e.g. with $q_1 = 4$ and $s_1 = 12$, $K_{SK} = 0.9976$.

Mutual inductance M is reduced by K_{SK} but the self-inductances L_1 and L_2 of stator and rotor remain unchanged. Alger [15] considers the skew-leakage inductance to be given by the change in $(L_1L_2 - M^2)/L_2$.

Hence $x_{SK} = (1 - K_{SK}^2)X_m$ 4.52

or, approximately, $x_{SK} = (1/12)(\pi k/s_1^2)X_m$. Half of x_{SK} may be apportioned to stator and rotor respectively.

It is noted that skew-leakage flux, greatest near the core ends, passes around the same magnetic circuit as the main flux. Skew leakage is therefore limited by magnetic saturation, particularly under conditions of high current, such as during starting.

Other authors [16] have given different procedures for incorporating the effect of skew in modified equivalent-circuit parameters. For example, the decoupling effect may be treated as a rotor winding factor, dividing all rotor impedances by K_{SK}^2 when referring them to the stator.

4.2.6 Peripheral Air-gap Leakage

Peripheral flux passing from pole to pole in the air-gap constitutes another component of leakage flux and, considering fundamental and phase-belt harmonics, the corresponding peripheral leakage reactance of either stator or rotor is

$$x_p = (2p^2g^2/D^2[1 + \Sigma(K_{pn}/nKp_1)^2]X_m, \ n \neq 1 \qquad 4.53$$

For a full-pitched, 3-phase, 60°-spread winding,

$$x_p = (2p^2g^2/D^2)\ (\pi^2/9)\ X_m \qquad 4.54$$

This is negligible for most induction motors and only becomes significant for machines with large ratios of air-gap/pole-pitch.

4.3 Referring Rotor Impedances to Stator

With a phase-wound rotor having m_2 phases, resistance and slot-leakage reactance x_{s2} at stator frequency are calculated, and then referred to equivalent values R_2' and x_2' per stator phase, using the impedance transformation ratio $(3/m_2)(NK_{w1}/N_2K_{wr})^2$ derived in Section 3.1.3.

With a cage rotor, resistance R_b (see Section 2.5.3) and slot-leakage reactance x_b are calculated per rotor bar. The transformation ratio is derived as follows. Equating fundamental mmf (equation 2.20) to its stator equivalent,

$$S_2 I_b / \sqrt{2}\pi p = (3/2)(4/\pi)(N/2p)K_{w1}\sqrt{2}\ I_2' \qquad 4.55$$

$$\therefore (I_2'/I_b) = S_2/6NK_{w1} \qquad 4.56$$

With fundamental flux per pole Φ_1,

$$E_b = (1/2)(4.44f\Phi_1)$$
$$\text{and } E_2' = 4.44fNK_{w1}\Phi_1$$

$$\therefore (E_2'/E_b) = 2NK_{w1} \qquad 4.57$$

$$\text{Hence } Z_2' = (E_2'/E_b)(I_b/I_2')Z_b$$

$$= 12(NK_{w1})^2(R_b + jx_b)/S_2 \qquad 4.58$$

The influence of skin effect upon R_b and x_b is considered in Chapter 6.

4.4 Summary of Leakage-Reactance Calculations

Bringing together the various components and subject to the qualifications already mentioned, the total leakage reactances of stator and rotor referred to the stator are respectively:

$$x_1 = (x_{e1} + x_{s1} + x_{Z1} + x_{B1} + x_{SK}/2 + x_{p1}) \qquad 4.59$$

$$x_2' = (x'_{e1} + x'_{s2} + x'_{Z2} + x'_{B2} + x_{SK}/2 + x'_{p2}) \qquad 4.60$$

The influence of magnetic saturation upon leakage reactances [17] must be mentioned. This is particularly important in respect of starting performance, since saturation of leakage reactances can cause a large increase of starting current above an unsaturated estimate.

Finally, a cautionary note is added concerning the temptation to substitute an "improved" formula for any individual component of leakage reactance. This could well make the overall result less accurate, as calculations of other components may well include empirical factors which compensate for errors in the component which is being corrected.

Plate 4. Cage rotor of a 4-pole, 320kW induction motor. The construction is for arduous high-speed applications and is designed to reduce stresses in operation and during pool brazing of end-rings.
Courtesy of Mather and Platt Machinery (UK)

CHAPTER 5
Relationships between Physical Design Features and Machine Parameters

5.1 Output Coefficient

One way of assessing the design of a conventional machine is to determine the output coefficient C, defined as

$$C = W/D^2L\omega_S \tag{5.1}$$

where W = air-gap power (W)
D = stator bore diameter (m)
L = stator core length (m)
ω_s = synchronous angular speed (rad/s)

The volume of the machine will be closely related to D^2L, and the torque at rated output will be related to W/ω_s. Hence C is a measure of rated torque per unit volume and the higher C the better utilised is the material of the machine.

Output coefficient is not the most important criterion of a design. Attention must be given to costs of materials and manufacture, temperature rises, efficiency, mechanical integrity and other application requirements.

Cost will, in general, increase with machine size although it is sometimes cheaper to use available components for an oversize machine rather than to use a more sophisticated or optimal design, requiring new tooling. In standardised ranges, it is common to find a range of outputs covered in the same frame size. Temperature rise will be determined by the losses and their distribution and by the effectiveness of the ventilation and cooling system. Efficiency is determined by the losses. It is possible that efficiency could be increased by raising flux and current densities, and hence output, but the increase in loss might well result in unacceptable temperature rises.

Mechanical integrity requires safe rotor stress levels. For a fixed geometry, mechanical stresses vary as (rotor peripheral speed)^2 or, for a given angular speed, as D^2.

Critical speed, or shaft whirling, is not usually a problem on medium and small machines, but unbalanced magnetic pull might be a problem for long machines (i.e. large L). However there is a significant advantage in using large values of D, rather than large values of L, to achieve a given rating.

5.2 Electric Loading

We have seen that three-phase stator currents produce an mmf which is virtually sinusoidal in its distribution with an amplitude given by

$$F_1 = (3\sqrt{2}/\pi)(NK_{w1}/p)\ I, \text{ ampere turns per pole} \qquad 5.2$$

If x is a distance around the periphery of the stator bore, the mmf distribution may be written as:

$$F = F_1 \sin(x\pi/\tau - \omega t) \qquad 5.3$$

This is produced by a set of axial currents in the stator winding which has an effective line density S A/m at the stator bore.

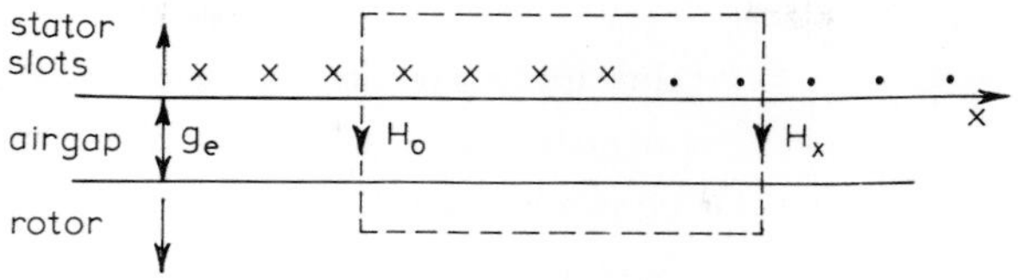

Fig. 5.1 Conditions at the airgap

Now $F = \oint H.dl = \Sigma i$ = total current enclosed by path of integration and $(H_x - H_o)g_e = F$ if H_o is chosen to be zero,

$$\text{with } \Sigma i = \int_0^x S dx$$

$$\therefore S = dF/dx = (\pi/\tau)F_1 \cos(x\pi/\tau - \omega t) = S\cos(\pi x/\tau - \omega t) \qquad 5.4$$

$$\text{where } S = (3\sqrt{2}/\tau p)NK_{w1}I = (6\sqrt{2}NK_{w1}/\pi D)I = (6NI/\pi D)\sqrt{2}K_{w1} \qquad 5.5$$

The total number of turns effectively carrying current I is 3N and total number of conductors is 6N.

$\therefore$ total slot r.m.s. current = 6NI

and r.m.s. current per metre of periphery $= 6NI/\pi D = A$ 5.6

A is called the electric loading of the machine and is an important design parameter. It is also equal to the total current (r.m.s.) in a stator slot divided by the slot pitch.

$S = \sqrt{2}K_{w1}A$ = peak fundamental line density (A/m) of stator currents 5.7

and $F_1 = (\tau/\pi)S = (D/\sqrt{2}p)K_{W1}A$
= amplitude of the armature mmf 5.8

A machine with a high value of electric loading A is one in which one or more of the following apply:

(i) the slot is deep - this increases slot-leakage inductance, tooth magnetising ampere turns and, with constant depth of core, increases the outside diameter.

(ii) the slot is wide with a narrow tooth - this reduces slot-leakage reactance but increases tooth flux density, thus increasing tooth magnetising ampere turns for a given flux per pole.

(iii) the current density in the copper is high - this increases the power loss (copper loss) per unit volume of the winding and requires improved cooling to avoid excessive temperature rise.

Note that $I = (\pi D/6N)A$ may be chosen as required, for a given A and D, by choice of N.

5.3 Magnetic Loading

The induced voltage per phase can be expressed as

$$E = (2\pi f/\sqrt{2})(NK_{w1})(\Phi_1) \tag{5.9}$$

With $\Phi_1 = (2/\pi)(\tau LB_1) = (DL/p)B_1$ and $\omega_s = 2\pi f/p$, this becomes

$$E = (1/\sqrt{2})NK_{w1}DL\omega_m(B_1/B_{gmax})B_{gmax} \tag{5.10}$$

where B_{gmax}, the maximum of the airgap flux density, is the magnetic loading of the machine.

The ratio B_1/B_{gmax} does not vary greatly and is always close to unity.

As has been shown in Chapter 3, the magnetic loading determines the magnetisation requirements of the machine and can only vary over a restricted range. A high value of B_{gmax} requires wide teeth, at the expense of narrow slots, and a deeper core depth if high flux densities are to be avoided. High flux density results in high magnetising mmf and high specific iron loss.

It can be seen that magnetic loading and electric loading compete for space in a given frame size. Generally an increase in one requires a reduction in the other. This gives rise to the comparison sometimes made between a "Copper" design and an "Iron" design.

In a "Copper" design, space has to be allocated to windings to give a relatively high value for electric loading A at the expense of magnetic loading B_{gmax}.

In an "Iron" design the opposite applies and, in view of the relatively high cost of copper, such a design is probably to be preferred.

The above expression shows that E may be chosen as required, for a given size of machine and B_{gmax}, by choice of N. If this is done, however, the relationship between I and A is also fixed.

5.4 C as a function of A and B_{gmax}

Let E and I be r.m.s.values of phase voltage and current at rated output with a power-factor angle φ.

The airgap power will be $3EI\cos\varphi$ and will be close to the output power W since the losses are generally small, and $\cos\varphi$ at the airgap will be similar to that at the terminals.

Hence,

$$W = 3EI\cos\varphi$$

$$= (3/\sqrt{2})NK_{w1}DL\omega_s(B_1/B_{gmax})B_{gmax}(\pi D/6N)A\cos\varphi \qquad 5.11$$

or

$$W = (\pi/2\sqrt{2})K_{w1}(B_1/B_{gmax})\cos\varphi AB_{gmax}D^2L\omega_s \qquad 5.12$$

Hence $$C = W/D^2L\omega_s = [(\pi/2\sqrt{2})K_{w1}(B_1/B_{gmax})(\cos\varphi)]AB_{gmax} \qquad 5.13$$

Of the terms in the square brackets which are not constants, $\cos\varphi$ is determined by either specification or other considerations, mainly associated with rotor windings, K_{w1} is usually close to 1.0 for all but some 2-pole machines, and B_1/B_{gmax} is close to 1.0. Thus C is a direct measure of how effectively the designer has utilised space to achieve the product AB_{gmax}.

A more informative coefficient might be

$$C' = \text{volt amperes}/(\text{rotor volume})\omega_s$$

$$= 4C/(\pi\cos\varphi)$$

$$= B_1(\sqrt{2}AK_{w1})$$

$$= B_1S$$

$$= (\text{peak fundamental gap flux density}) \times (\text{peak fundamental electric loading}) \qquad 5.14$$

The above illustrates the importance of A and B_{gmax} in a given machine frame. The choice of turns/phase N affects only the impedance of the machine, or ratio of voltage to current, for given electric and magnetic loadings.

The number of stator slots might seem to be equally unimportant, but this is not so. The larger the number of slots, the better the emf and mmf waveforms. There is, however, an increased loss of space owing to an increased number of slot insulation thicknesses.

5.5 Effects of Size upon C

In order to reduce the number of parameters being varied, let us assume a machine cross-section of constant geometry (with constant number of slots, poles and constant winding arrangement) but variable in its scale, i.e. with D and/or L variable.

As D varies, let us keep constant all flux densities in the iron and all current densities in the copper. This has the effect of keeping specific loss (loss per unit volume) constant at all corresponding points. The symbol α will be used to denote proportionality.

If flux densities are constant, then B_{gmax} is constant.

The cross-sectional area of a conductor α D^2. Thus, current per slot, for constant current density J, α D^2 but slot pitch α D.

Therefore electric loading = (current/slot)/(slot pitch) = A α D

Hence,

$$\text{Output Coefficient } \alpha\ B_{gmax}A\ \alpha\ D. \qquad 5.15$$

So, the bigger the diameter, the better is the material utilisation for fixed densities. Unfortunately for large-machine designers (or, fortunately for small-machine designers) a problem arises if J is kept constant.

Consider a particular type of winding and assume that it has a constant thickness of insulation irrespective of scale. In practice the number of turns of the winding would be adjusted to maintain a required terminal voltage as the scale of trial designs was changed. We will assume that all currents and voltages are referred to a constant number of turns. Constant insulation thickness will enhance the effect of increase of A with D, but thermal effects cannot be ignored. There will be a thermal resistance per unit length of machines R_{th} between copper and ambient and most of this will be due to the air and electrical insulation in the slot. If the thickness of insulation is constant then R_{th} will decrease linearly with scale, or

$$R_{th} \propto D^{-1} \qquad 5.16$$

The power loss in the copper, per unit length, will vary approximately quadratically with scale for a given J, or

$$\text{loss} \propto J^2D^2 \qquad 5.17$$

The temperature rise $(\Delta\theta) = (\text{loss})R_{th}$, so

$$\Delta\theta \propto J^2D$$

or, for a given temperature rise,

$$J \propto D^{-\frac{1}{2}} \qquad 5.18$$

This explains why small machines can tolerate higher current density in their conductors than large machines. The effect of this is that

$$A \propto D^{\frac{1}{2}} \qquad 5.19$$

The effect of scale upon temperature rise caused by iron loss is not so pronounced, since the iron is in direct contact with the cooling medium and it is relatively easy to arrange for improved core cooling as size increases. Thus magnetic loading does not vary much with scale on account of thermal considerations.

The consequence of the above effects is that

$$C \propto D^{\frac{1}{2}}. \qquad 5.20$$

This is still an improvement in utilisation as diameter increases.

For very large machines with direct conductor cooling and heavy insulation, A can be as high as 200 A/mm (2×10^5 A/m or 5000 A/in).

For very large machines with indirect cooling A may be of the order of 80 A/mm (2000 A/in) and for small machines, with mush windings, A may be of the order of 30 A/mm (750 A/in).

5.6 Effects of Size upon X_m

Let us assume B_{gmax} is constant and $J \propto D^{-\frac{1}{2}}$ as the scale of our machine cross-section changes.

As D increases, magnetising mmf = $\oint \mathbf{H}.dl$ increases linearly with D since H remains constant at all corresponding points. Thus, for a given B_{gmax} and winding layout,

$$I_m \propto D \qquad 5.21$$

Flux/pole $\propto$ DL, so E $\propto$ DL and I $\propto$ JD^2 or I $\propto$ $D^{3/2}$

so that

$$X_m = (E/I_m) \propto L$$

and

$$X_m \text{ per-unit} = X_m(I/E) \propto D^{\frac{1}{2}} \qquad 5.22$$

So X_m (per-unit) increases with increase in diameter. The length of the airgap of an induction motor is made as small as possible, the minimum being determined by mechanical clearance, unbalanced magnetic pull, and parasitic surface losses caused by the slot openings. Consequently, as size increases, the airgap does not increase proportionally. This has the effect of increasing X_m even more as size increases.

Thus, for a given geometry, the bigger the diameter the bigger is per-unit X_m or the smaller is the per-unit magnetising current I_m. This increases $\cos\varphi$ for an induction motor, and is an additional benefit in terms of output coefficient. Table 5.1 illustrates the effect of size upon magnetising requirements for some typical 4-pole induction motors.

Table 5.1 MAGNETISING REQUIREMENTS OF INDUCTION MOTORS

Rated output, kW	0.1	0.5	2.5	15	600
I_m, per-unit	0.78	0.52	0.5	0.3	0.2
$\cos\varphi$ at full load	0.7	0.75	0.8	0.84	0.9

It should be noted that, for given values of D, L, A and B_{gmax}, the per-unit value of X_m decreases as p increases, or the pole-pitch reduces. This is shown as follows:

X_m per-unit $= (E/I_m)(I/V)$ where I and V are rated values.

Now $E \simeq V$, so that X_m per-unit $\simeq I/I_m = 1/(\text{per-unit } I_m)$

$$I = (\pi D/6N)A$$

$$\text{Magnetising mmf} = F_m = F_{gap}\{1 + (2/\sqrt{3})F_{iron}/F_{gap})\}$$

and $I_m = F_m \pi p/3\sqrt{2}NK_{w1}$

F_m will not vary much if B_{gmax} and g_e are constant as p, and hence the pole-pitch τ, varies. So,

$$X_m \simeq I/I_m = \pi DA3\sqrt{2}NK_{wl}/6NF_m\pi p = DAK_{wl}/\sqrt{2}F_m p \qquad 5.23$$

Thus, per-unit magnetising reactance is universely proportional to p for given size and electric and magnetic loadings. For small machines, F_m must be reduced, by reducing B_{gmax}, for large p to make the above ratio exceed unity. If this were not done, then the magnetisation requirements would result in a current greater than the rated current corresponding to the electric loading A.

5.7 Effects of Size upon Leakage Reactance

Within the axial length of most conventional machines, the cross-slot leakage flux is responsible for a substantial component of winding leakage reactance. If the geometry and winding layout are constant, the cross-slot leakage permeance will be constant as D varies, as slot depth and width dimensions vary in the same ratio, but will vary linearly with L.

Assuming current density $J \alpha D^{-\frac{1}{2}}$, then

$$\text{current/slot} \ \alpha \ D^{3/2}L$$

and leakage-flux linkage/slot α $D^{3/2}L$.

If x_s is the cross-slot component of leakage reactance, the corresponding reactance voltage drop will be Ix_s, proportional to leakage flux linkage per slot. So,

$$Ix_s \ \alpha \ D^{3/2}L$$

The generated voltage is E, proportional to DL for given B_{gmax}, and hence:

$$\text{per-unit } x_s = Ix_s/E \propto D^{\frac{1}{2}} \qquad 5.24$$

Thus, per-unit leakage reactance increases with size.

5.8 Variation of X_m and x_s

The designer can only control the per-unit X_m by changing the ratio of armature reaction mmf to magnetising mmf. This can be done by changing A or B_{gmax} or the airgap length. A reduction in electric loading A will effectively reduce per-unit X_m but will also reduce C and the rated output for a given frame.

An increase in B_{gmax} will increase the iron loss (for a given geometry), so the usual step taken is to increase the airgap length if it is required to decrease X_m per-unit (i.e. to increase the Short-Circuit Ratio).

The leakage reactance is most directly affected by the slot geometry. Changing the number of slots has little effect upon x_s if the ratio slot-width/slot-pitch is maintained for given B_{gmax} and A.

The per-unit value of x_s can be reduced by reducing the slot-depth with a consequent reduction in A, or increasing the slot-width/slot-depth ratio (with a consequent increase in A and reduction in B_{gmax}) or a combination of these. Note that both will result in a reduction in overall machine diameter.

The required per-unit value of leakage reactance is thus an important parameter having a significant effect upon the machine design. For induction motors the leakage reactance determines starting current and maximum torque and is therefore obviously an important parameter in machine performance. For synchronous machines the leakage reactance is less important for normal synchronous operation. However, for motors, it is almost as important as for the induction motor if the motor is to be started asynchronously on load. For generators, it is a principal component of the subtransient reactance which limits fault currents.

5.9 Effects of Size upon Resistance and Losses

Let us assume, still, constant geometry and winding layout as scale changes, with B_{gmax} constant and $J \propto D^{-\frac{1}{2}}$.

We have seen that, with these assumptions, the permissible loss in the copper of a winding per unit length varies as J^2D^2 or, since $J \propto D^{-\frac{1}{2}}$, varies as D. Thus, total loss

$$I^2R \propto DL. \qquad 5.25$$

R per-unit = R(I/V) where I and V are rated values

$$= (I^2R/VI) \simeq (I^2R/EI) = (\text{ohmic loss,W})/(\text{rated VA})$$

Now we have also seen that $E \propto DL$ and $I \propto D^{3/2}$.

Hence

$$R \text{ per-unit, and per-unit copper loss, } \propto D^{-3/2} \qquad 5.26$$

This shows that, the greater the diameter, the smaller is the full load ohmic loss as a fraction of the output.

For constant flux densities, the iron loss $\propto D^2L$.

For given cooling medium, and at fixed speed, the mechanical loss $\propto D^3L$ approximately.

Total loss = copper loss + iron loss + mechanical loss, which could be written as :

$$\text{loss} = aDL + bD^2L + cD^3L$$

and the output as

$$\text{output} = dD^{5/2}L$$

where a, b, c and d are constants.

So, dividing loss by output, per-unit iron loss α $D^{-\frac{1}{2}}$ and per-unit mechanical loss α $D^{\frac{1}{2}}$.

Efficiency = output/(output + loss)

$$= dD^{5/2}L/(dD^{5/2}L + adL + bD^2L + cD^3L) \qquad 5.27$$

Differentiation of this expression with respect to D shows that efficiency increases as D increases provided that:

(copper loss + iron loss) > (mechanical loss/3)

which is usually the case.

The foregoing has indicated that, in the majority of cases, significant advantages accrue from designing for large D and small L, provided that end-effects do not become dominant.

There is obviously a limit to the D/L ratio, beyond which neglect of end-region resistance, losses, fluxes and cooling becomes serious and invalidates some of the previous analyses. It is not possible to incorporate end-effects into a general treatment.

A machine designer is usually conscious of machine losses because of their heating effects, and probably to a lesser degree because of their effect upon efficiency. Rarely will he include, as a most significant criterion in his choice of design parameters, the subdivision of losses between iron and copper losses.

Consider a chosen frame size and geometry and, for a given output, change the rated values of V and I by changing B_{gmax} and A, remembering that a particular voltage can be obtained by adjusting N. In general, iron losses P_i in the magnetic circuit will vary with voltage, whereas losses in conductors P_c, that is the load loss, will vary with current and mechanical loss P_m will remain constant.

$$\text{Let } P_c = aI^m$$
$$P_i = bV^n$$
$$P_m = c$$

where a, b, c, m and n are constants.

The output P is proportional to VI, and if P is kept constant then V = P/I as V and I are varied.

$$\text{Efficiency} = \eta = P/(P + P_c + P_i + P_m)$$

or

$$\eta = P/(P + aI^m + bP^nI^{-n} + c)$$

Differentiating η with respect to I shows that a maximum occurs when $amI^{m-1} = nbP^nI^{-(n+1)}$, that is when

$$mP_c = nP_i \qquad 5.28$$

Thus, maximum efficiency is obtained when

$$\text{iron loss} = (m/n)(\text{load loss})$$

The index m is usually close to 2, since the majority of the load loss is an ohmic loss in the conductors. The index n can also be close to 2, but in highly saturated conditions could be higher.

Power transformers are designed so that the maximum efficiency occurs at part rated load, where they operate for the longest periods. Machine design is subject to more constraints than transformer design, so that it is a fortunate designer who finds that his machine has maximum efficiency at rated conditions. Often, however, these two points are not greatly separated.

Plate 5. Stator of a 1485 MW, 20kV, 4-pole turbine generator. Weight = 183 tonne. Full-load efficiency = 99.06%
Courtesy of GEC Alsthom Turbine Generators Limited

CHAPTER 6
Eddy Currents and Deep-Bar Effects

6.1 General

Time-varying leakage fluxes induce emfs and eddy currents in conducting parts, giving rise to eddy-current losses. Thus, stator-slot leakage fluxes cause eddy currents in stator conductors. These eddy currents flow in a direction opposing the change of flux, adding to the slot current at the top of the slot and subtracting from it at the bottom. Expressed another way, the lower conductor elements are linked by most of the slot leakage flux and therefore have higher leakage reactance, causing their currents to be smaller and more lagging in time phase than the currents in conductor elements near the top of the slot. The result is a non-uniform distribution of current and a consequent effective increase of winding resistance. The ratio of a.c. to d.c. resistance of the slot portion of a stator winding increases sharply with strip depth so, for large machines, it is common to subdivide deep conductors into separately-insulated strips in the slot depth. Transposition of strip position, such that each strip occupies all positions in the slot for an equal slot length, is also used to cancel emfs induced by leakage fluxes. Strips may then be connected in parallel without causing circulating currents. In very large machines, attention is also given to end-leakage fluxes in this respect.

In this Chapter some basic theory of eddy currents is presented and a common application, in deep-bar induction motors, is then treated.

6.2 Field not Modified by Eddy Currents

When considering problems involving induced effects produced in conducting components by time-varying fields, it is important to distinguish between applied fields which remain unmodified and those which are modified by the induced eddy currents.

When the eddy-current field is relatively weak, so that the externally applied field remains unchanged throughout the material under consideration, the calculation of the induced loss is relatively simple.
Fig. 6.1 shows a rectangular slab of material situated in a region subject to a sinusoidal variation of flux density $B = B_m \sin\omega t$. The surface area of one face of the slab is LD and its depth, normal to the direction of the incident flux, is 2b. Eddy currents flow in the direction shown, tending to oppose the change of flux.

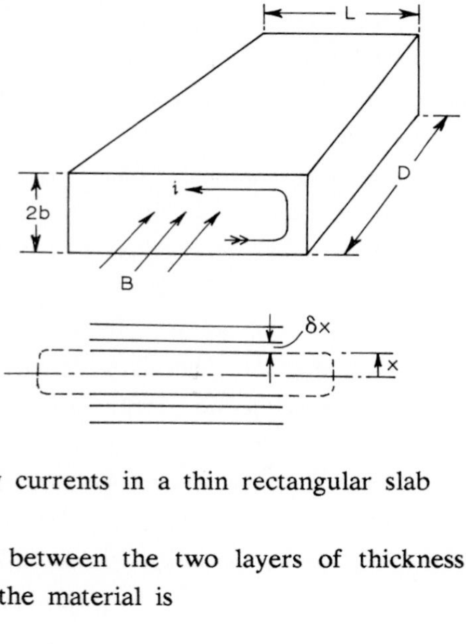

Fig. 6.1. Eddy currents in a thin rectangular slab

The flux enclosed between the two layers of thickness δx at distance x from the centre of the material is

$$\Phi = 2LxB_m \sin\omega t \qquad 6.1$$

The r.m.s. value of the emf induced in the circuit is:

$$E = (1/\sqrt{2})(d\Phi/dt)$$

$$= \sqrt{2}(Lx\omega B_m) \qquad 6.2$$

Neglecting the end-closing paths, the resistance of the circuit is $2\rho L/D\delta x$ and the element of eddy-current loss is therefore $E^2D\delta x/2\rho L$ or $2(Lx\omega B_m)^2D\delta x/2\rho L$. Hence, the total eddy-current loss is:

$$W = \int_0^b (Lx\omega B_m)^2(D/\rho L)dx$$

$$= (b^3/3\rho)B_m{}^2\omega^2 LD \qquad 6.3$$

This result shows that, in this case, the loss is inversely proportional to the resistivity of the material and proportional to the cube of the depth of the slab measured normal to the direction of the flux. Machine and transformer cores are therefore constructed of thin laminations of high-resistance coreplate material in order to minimise the loss.

In the type of problem considered above, the eddy currents are said to be "resistance limited", for obvious reasons. When the strength of the eddy currents is sufficient to modify both the magnitude and phase of the flux as a function of depth into the material, there is said to be appreciable skin effect and a more refined analysis is required.

The simple analysis described in this Section may only be applied when the half-depth b of the conducting material is much less than the depth of penetration δ described in Section 6.3.

6.3 Field Modified by Eddy Currents

The classical theory of eddy-current effects [18] applies to conducting materials having constant permeability, such as copper, aluminium and good-quality non-magnetic steel.

Consider a plane-faced block of conducting material to which a uniform sinusoidal magnetising force $H_m \sin\omega t$ is applied, as shown in Fig. 6.2. Let the r.m.s. value of the magnetising force ($H_m/\surd 2$) be called H_o.

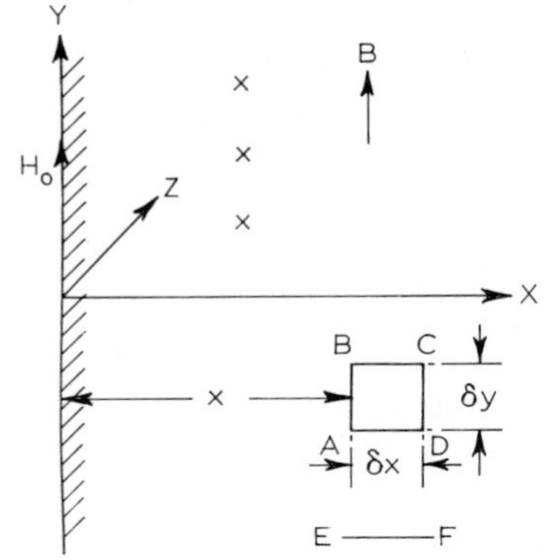

Fig. 6.2. Section of a plane-faced block

The resultant magnetising force H at any point at a depth x within the block is the resultant of H_o and the field due to the eddy currents.

The corresponding flux density B and current density J are also functions of x.

Applying the magnetic-circuit law to the elemental rectangular path ABCD shown,

$$H\delta y - (H + \delta H)\delta y = J\delta y\delta x$$

and, in the limit,

$$dH/dx = - J \qquad 6.4$$

Considering a circuit which is infinitely long in the z direction and whose end view is EF, the current densities are J and (J + δJ) at E and F respectively and the corresponding electric forces are ρJ and ρ(J + δJ).

The difference between these gives the emf induced by the change in the flux linking the circuit. Thus

$$-\rho JL + \rho(J + \delta J)L = -d\Phi/dt$$

$$= -j\omega\Phi$$

$$= -j\omega BL\delta x$$

Hence

$$\rho dJ/dx = -j\omega B \qquad 6.5$$

We also have the relationship

$$B = \mu H$$

where $\mu = \mu_r\mu_o$ 6.6

The solution for flux density is obtained as follows.

From eqn. 6.5, $B = j(\rho/\omega)dJ/dx$

and substituting $J = - dH/dx$ from eqn. 6.4,

$$B = - (j\rho/\omega)d^2H/dx^2$$

Using $B = \mu H$,

$$B = - j(\rho/\mu\omega)d^2B/dx^2$$

Or

$$d^2B/dx^2 - j(\mu\omega/\rho)B = 0 \qquad 6.7$$

Putting

$$(\rho/\mu\omega)^{\frac{1}{2}} = \delta \qquad 6.8$$

$$d^2B/dx^2 - (j/\delta^2)B = 0 \qquad 6.9$$

The solution to this equation for B is

$$B = B_o e^{-x(1+j)/(\delta\sqrt{2})} \qquad 6.10$$

$$= B_o e^{-x/\delta\sqrt{2}} e^{-jx/\delta\sqrt{2}}$$

$$= B_o e^{-x/\delta\sqrt{2}} \{\cos(x/\delta\sqrt{2}) + j\sin(x/\delta\sqrt{2})\} \qquad 6.11$$

where $B_o = \mu H_o$ is the value of B at x = 0, i.e. the r.m.s. value of flux density at the surface of the block.

The magnitude of the flux density, $B_o e^{-x/\delta\sqrt{2}}$, is thus seen to decrease exponentially with the increase of distance x into the block. The quantity δ, called the depth of penetration, provides a comparative measure of the extent of flux penetration. Note that some authors define a depth of penetration which is $\sqrt{2}$ times larger than that given by eqn. 6.8.

e.g. (i) for copper at 50 Hz, $\mu_r = 1$, $\rho = 1.7 \times 10^{-8}$ Ωm and $\delta = 6.6$ mm

(ii) for iron at 50 Hz, with $\mu_r = 1000$ and $\delta = 10^{-7}$ Ωm, $\delta = 0.5$ mm

Maximum useful thicknesses of busbars and core laminations at 50 Hz are indicated by (i) and (ii).

To solve for current density, from equations 6.4 and 6.6,

$$J = -(1/\mu)dB/dx$$

Using the expression $B_o e^{-x(1+j)/\delta\sqrt{2}}$ for B, from equation 6.10,

$$J = (B_o/\mu)(1+j)/\delta\surd 2)e^{-x(1+j/\delta\surd 2)}$$

$$= J_o e^{-x(1+j)/\delta\surd 2} \qquad 6.12$$

where $J_o = (B_o/\mu\delta)(1+j)/\surd 2$

and $|J_o| = H_o/\delta$ 6.13

The current density J thus varies exponentially with x, in the same way as B. The non-uniform distribution of current and flux densities, with high values near the surface of the material, is described as "skin effect" and leads to higher values of eddy-current loss than would be obtained with uniform distribution.

The total eddy-current loss is the integral of the loss in the elemental strip of length L in the current direction z, at depth x.

Thus,

$$\text{loss} = \int_o^\infty (\rho L/Y\delta x)(|J|Y\delta x)^2$$

$$= \int_o^\infty \rho|J|^2 LYdx$$

$$= \int_o^\infty \rho J_o^2 e^{-x\surd 2/\delta} .dx, \text{ per unit surface area}$$

$$= (\rho\delta/\surd 2)J_o^2, \text{ per unit surface area} \qquad 6.14$$

If the applied magnetising force H_o is known, an alternative form of eqn. 6.14 in terms of H_o is more useful. Substituting for J_o from eqn. 6.13 and for δ from eqn. 6.8,

$$\text{loss} = \surd(\rho\mu\omega/2)H_o^2 \qquad 6.15$$

In contrast to the "resistance limited" case, the eddy-current loss is proportional to the square root of the resistivity when there is appreciable skin effect. This means that, in order to minimise losses under these conditions, the resistivity should be made as low as possible and the component should be sufficiently deep to allow unrestricted flow of the eddy currents. That is, the depth should be appreciably greater than the depth of penetration δ. This is the basis of design for eddy-current screens.

Equation 6.15 shows that the loss is proportional to the square root of the permeability. This is an important consideration in the design of non-magnetic steel components which are subject to appreciable alternating fields. If poor-quality non-magnetic steel is used, the relative permeability may be as high as 20. The eddy-current loss would then be 4-5 times the value estimated on the basis of unity permeability.

The total flux per unit length in the z direction is

$$\Phi = \int_0^\infty B\,dx$$

$$= \int_0^\infty B_0 e^{-x(1+j)/\delta\sqrt{2}}\,.dx$$

$$= (B_0\delta)\sqrt{2}/(1+j) \qquad 6.16$$

The magnitude of flux $(B_0\delta)$ per unit length is equivalent to that of a uniform flux density B_0 extending over a depth δ into the material. The second bracket in eqn. 6.16 shows that the total flux in the material lags the surface flux density by 45°.

It is often found convenient in machine analysis to use effective lumped circuit impedances to represent distributed effects and, in the case of eddy-current losses, it is then necessary to know the equivalent phase angle as well as the total power loss. This may be deduced as follows:

Surface impedance per unit area = $\bar{Z} = \bar{E}/H_o$

In the present example,

$$\bar{E} = j\omega\Phi = j\omega B_o\delta\sqrt{2}/(1+j)$$
$$= \omega B_o\delta\angle 45^o$$

and
$$\bar{Z} = \bar{E}/H_o = \omega(B_o/H_o)\delta\angle 45^o$$
$$= (\mu\omega\delta)\angle 45^o$$
$$= (\mu\omega\rho)^{\frac{1}{2}}\angle 45^o$$

The phase angle of the effective surface impedance is 45°. That is, the effective resistance and reactance are equal.

This treatment is for a slab of infinite depth. For a finite thickness, a similar analysis can be made taking account of the boundary conditions to give a modified expression for $\bar{E}/H_o$ at the surface.

For the case of conductors in slots, the assumption that the flux density is directed in straight lines across the slot-width enables a single slot to be treated as part of an infinite system, as shown below.

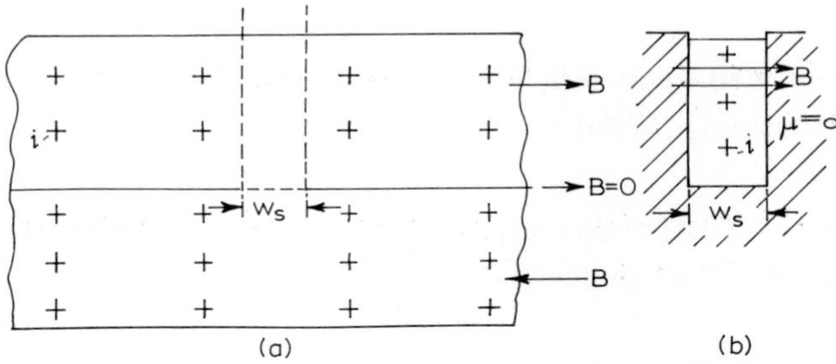

Fig. 6.3. Illustration of a conductor in a parallel-sided slot.
(a) equivalent system (b) single slot

6.4 Deep-bar Rotor

Deep-bar cage rotors are designed to exhibit substantial skin effect, such that their secondary resistance is large at high slip frequency and decreases as slip decreases.

For a simple rectangular bar of width w_b, depth d and resistivity ρ, in a slot of width w_s, effective resistivity is equal to $\rho w_s/w_b$. The complex bar impedance [19] including skin effect is

$$\bar{Z}_b = (L\rho/w_b)(\sqrt{j}/\delta)\coth\sqrt{j}(d/\delta) \qquad 6.17$$

where $$\delta = [(\rho/\mu_o)(w_s/w_b)/(2\pi f)]^{\frac{1}{2}} \qquad 6.18$$

This may be developed to give the components of $\bar{Z}_b$ as

$$R_b = R_{dc}(\sqrt{2}d/2\delta)[\sinh(\sqrt{2}d/\delta) + \sin(\sqrt{2}d/\delta]/ \\ [\cosh(\sqrt{2}d/\delta) - \cos(\sqrt{2}d/\delta)] \qquad 6.19$$

$$x_b = R_{dc}(\sqrt{2}d/2\delta)[\sinh(\sqrt{2}d/\delta) - \sin(\sqrt{2}d/\delta/)/ \\ [\cosh(\sqrt{2}d/\delta) - \cos(\sqrt{2}d/\delta)] \qquad 6.20$$

where $$R_{dc} = (L\rho/w_b d) \qquad 6.21$$

The influence of size upon skin effect is expressed by the ratio $(\sqrt{2}d/2\delta)$ or, with a given material, by $d\sqrt{f}$.

If $\sqrt{2}(d/2\delta) > 2$ (i.e. large skin effect), R_b and x_b both become equal to $\sqrt{2}(d/2\delta)R_{dc}$ and the phase angle is 45°.

The ratio $K_b = R_b/R_{dc}$, given by eqn. 6.18, is the increase of bar resistance owing to skin effect. As calculated in Section 6.3, δ for copper at 50 Hz is 6.6mm. Then to achieve $K_b = 2.5$, for example, the required depth of bar is 23.3 mm.

Slot leakage permeance or inductance is reduced by skin effect and is evaluated using eqn. 6.19.

6.5 Non-rectangular Bar

A special procedure has been developed for computation of the impedance of a non-rectangular rotor bar, including skin effect. Such a bar is subdivided into a number of sections, each of which is rectangular. A trapezoidal section, such as occurs with a tapered bar, may be subdivided into a large number of rectangular strips, each represented by its depth and average width. Computation commences with the bottom section, furthest from the air-gap, and calculates the impedance of this section using eqn. 6.16 above.

Progressing section-by-section up the bar, the total impedance is obtained using the following relationship [20],

$$\bar{Z}_n = [\bar{A}\cosh\sqrt{j}(d/\delta) + (\bar{A}^2/\bar{Z}_{n-1})\sinh\sqrt{j}(d/\delta)]/[\sinh\sqrt{j}(d/\delta) + (\bar{A}/\bar{Z}_{n-1})\cosh\sqrt{j}(d/\delta)] \qquad 6.22$$

where $\bar{Z}_n$ = impedance seen at top of nth section
$\bar{Z}_{n-1}$ = impedance seen at the bottom of nth section
$\bar{A} = (L\rho/w_b\delta)\sqrt{j}$
$= R_{dc}\sqrt{j}(d/\delta)$ of the nth section

For the bottom section, or a single rectangular section,

$$\bar{Z}_{n-1} = Z_o = \infty$$

and then

$$\bar{Z}_b = \bar{A}\coth\sqrt{j}(d/\delta) \qquad 6.23$$

as before.

Skin-effect factors for tapered and rectangular (b=1) bars may be evaluated using Figs. 6.4–6.6, for which δ is to be calculated using w_b = mean bar-width and w_s = mean slot-width.

Values of R_b and x_b including skin effect should be incorporated, where appropriate, in the calculation of rotor parameters described in Chapter 4.

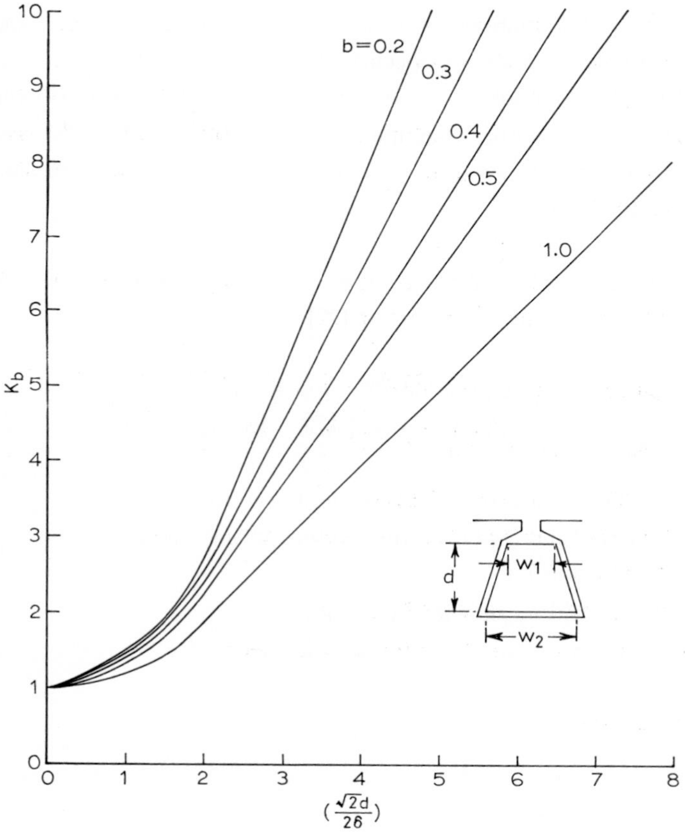

Fig. 6.4. Resistance-increase factor K_b

$K_b = R_b/R_{dc}$

$b = w_1/w_2$

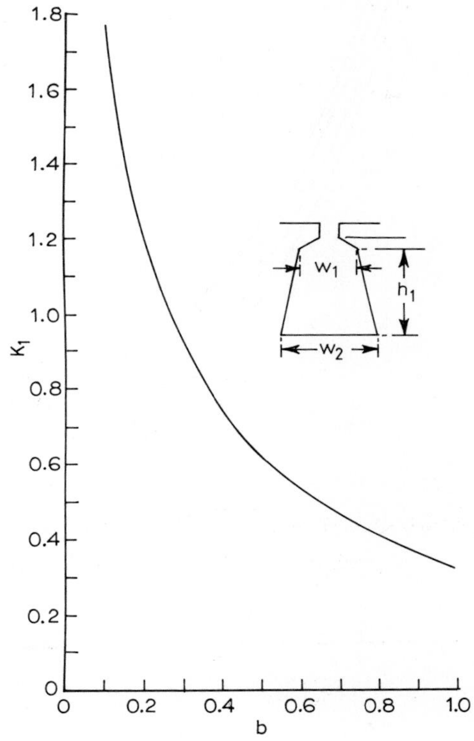

Fig. 6.5. Slot-permeance factor K_1
Slot permeance = $(h_1/w_2)K_1K_2$
$b = w_1/w_2$
K_2 obtained from Fig. 6.6

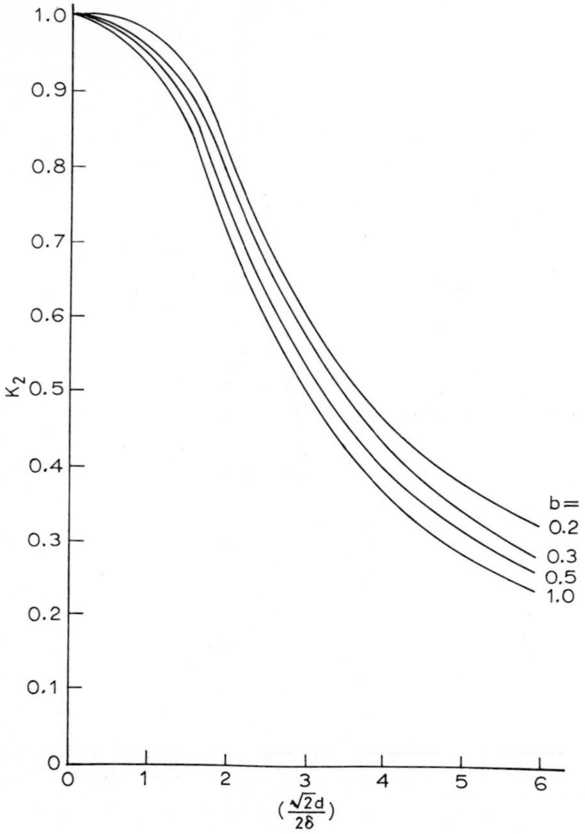

Fig. 6.6. Slot-permeance factor K_2 for use with Fig. 6.5
$b = w_1/w_2$, see Fig. 6.5

Plate 6. Deep–bar rotor of a 2.9 MW, 4–pole cage induction motor for a turbo–compressor drive. The ratio of rotor resistances is 3.7 between standstill and running conditions. Starting current = 5.3 times full–load current, starting torque = 0.73 times full–load torque

Courtesy of Brush Electrical Machines Limited

CHAPTER 7
Transient Reactances and Time Constants

7.1 Machine Transients

The conventional equivalent circuit, or phasor diagram, of an a.c. machine cannot be used to represent any form of transient. It happens that in the steady-state, for most three-phase machines, the airgap flux distribution is almost constant. Then all stator phases experience balanced conditions and the rotor and its circuits can be represented by an equivalent set of windings also carrying balanced currents.

For any form of transient, the balance does not exist and a more generalised treatment of the variation of inductances with time must be used. The most widely-used method is the two-axis dynamic theory which treats air-gap fluxes as having two components along d- and q-axes, which can be identified with axes of magnetic symmetry in the case of machines which have saliency on one member. The theory cannot be used for machines which have saliency or unbalanced windings on both members. It makes approximations with regard to the variation of inductances with rotor position and cannot include precisely the effect of magnetic saturation. Nevertheless it is widely used, with success, to predict the behaviour of machines during transient conditions.

When both stator and rotor windings are replaced by d- and q-axis equivalents, a transformation is used which eliminates the dependence of inductances upon rotor position. This transformation replaces a three-phase winding by equal coils on each axis, carrying transformed currents. The d-axis coil carries current i_d and is linked by d-axis flux, to give a flux linkage λ_d, and the q-axis coil has current i_q with flux linkage λ_q.

The axis voltages are then given by:

$$v_d = i_d R_d + d\lambda_d/dt + \lambda_q(d\theta/dt) \qquad 7.1$$

$$v_q = i_q R_q + d\lambda_q/dt + \lambda_d(d\theta/dt) \qquad 7.2$$

where θ is a rotor angle.

There are three types of voltage in each expression - ohmic, transformer and rotational. Usually $R_d = R_q$. The most convenient transformation results in the inductances and resistances being the same as the per-phase values of the three-phase winding.

For synchronous machines it is essential that the d- and q-axes be fixed to the rotor and they are usually taken as the same as those described in the steady-state treatment. Hence the d-axis flux is produced only by d-axis mmfs, which may be produced by the d-axis current of the stator, the field current and the currents in damper circuits on the d-axis of the rotor. The q-axis flux is produced by the q-axis current of the stator and currents in damper circuits on the q-axis of the rotor. There is no transformer coupling between any two coils on different axes.

In the steady-state, the above transformation results in d- and q-axis stator currents which are constant, flux linkages λ_d and λ_q which are also constant, and rotor currents which are constant or zero. The d- and q-axis stator voltages are then ohmic and rotational. These constant quantities transform into alternating phase voltages and currents at supply frequency.

7.2 Equivalent Circuits

During a transient, the axis currents change, giving rise to transformer-type voltages which can be regarded as acting on the equivalent circuits shown in Fig. 7.1

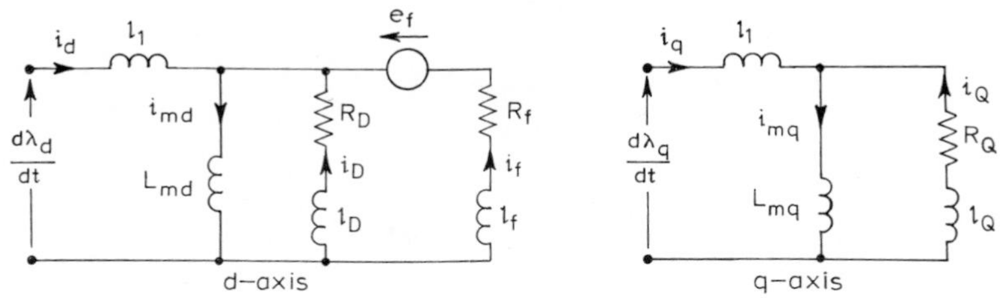

Fig. 7.1. Equivalent circuits for transformer-type voltages on d- and q-axes

In Fig. 7.1.,

L_{md} = d-axis magnetising inductance
L_{mq} = q-axis magnetising inductance
l_1 = stator leakage inductance
R_f = field resistance
l_f = field leakage inductance
R_D = d-axis damper resistance
l_D = d-axis damper leakage inductance
R_Q = q-axis damper resistance
l_Q = q-axis leakage inductance

and all are referred to the stator per phase.

These equivalent circuits are convenient for identification of some of the reactances and time constants used in synchronous-machine analysis.

A well known and common test made on a synchronous machine is a solid three-phase short-circuit at its terminals from open-circuit, when excited at a suitable voltage. A record of the ensuing variation of winding currents enables various reactances and time constants to be deduced.

Figs. 7.2 and 7.3 show typical curves of winding currents after such a short-circuit on a large machine.

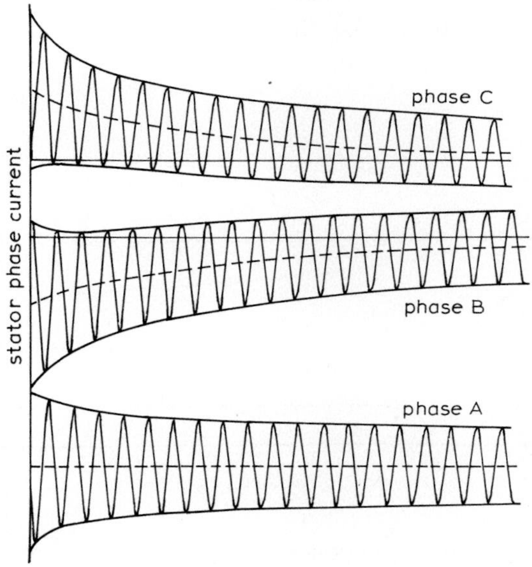

Fig. 7.2. Oscillograms of armature currents after a short-circuit

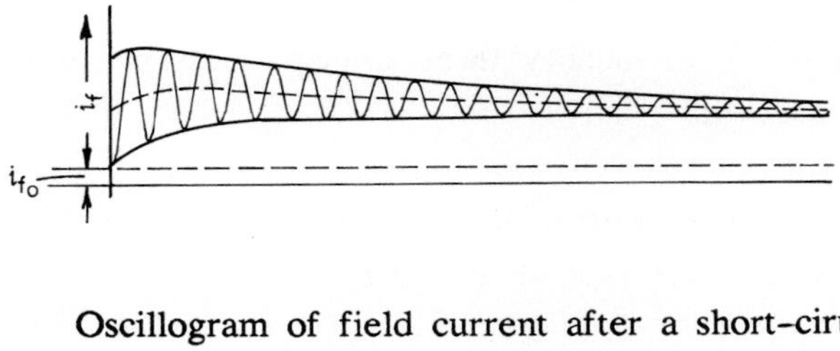

Fig. 7.3. Oscillogram of field current after a short-ciruit

Envelope lines have been drawn through the peaks of the alternating currents, and dotted lines drawn through the mean of the envelope lines. The dotted lines represent asymmetrical components of current, or transient unidirectional (d.c.) currents, to which the transient alternating components are added. Both d.c. and a.c. components start with an initial value, which may be determined by extending the envelope curves back to zero time, and decay in a complex fashion.

If the a.c. components from Fig. 7.2 are converted to r.m.s. or per-unit values and plotted against time, the same curve should be obtained for all three phases and will be of the form shown by the full-line AB in Fig. 7.4.

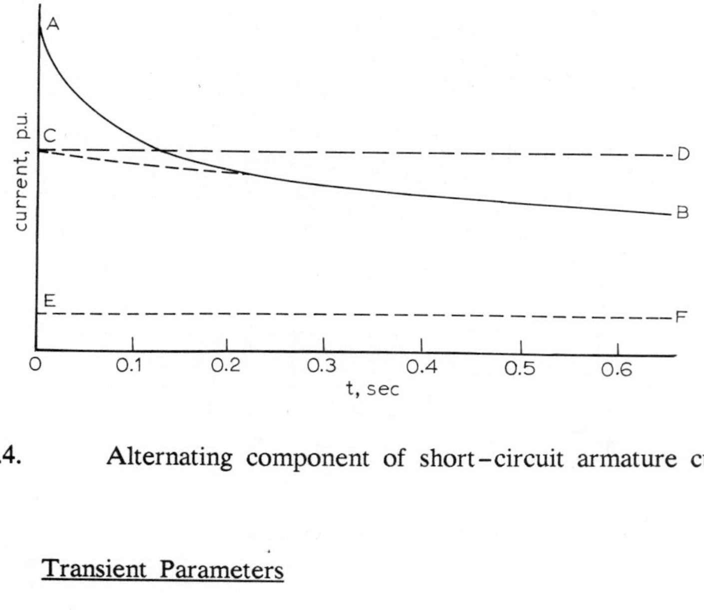

Fig. 7.4. Alternating component of short-circuit armature current

7.3 Transient Parameters

It is usually possible to identify three principal sections of the curve AB in Fig. 7.4, as follows:

(i) a steady-state short-circuit current, reached when all transients have ended, corresponding to EF

(ii) a region of exponential decay with a relatively large time constant, which can be extended to zero time to give the curve CB. The difference between CB and EF is called the transient component

(iii) a region of exponential decay with a short time constant, corresponding to the difference between AB and CB. This is called the sub-transient component.

If V is the open-circuit phase voltage before short-circuit, this will correspond to $X_{md}I_e$, where I_e is the field current referred to the stator, r.m.s. per phase, and the current OE = V/X_d where X_d is the d-axis synchronous reactance (cf. Section 3.2.2).

The "initial" current OC = V/x_d', where x_d' is the d-axis transient reactance

The current OA = V/x_d'', where x_d'' is the d-axis subtransient reactance.

The time constant of the decay of the difference between AB and CB (the sub-transient period) is called the d-axis short-circuit sub-transient time constant T_d''.

The time constant of the decay of the difference between CB and EF (the transient period) is called the d-axis short-circuit transient time constant T_d'.

The behaviour just described can be explained by the axis equivalent circuits and the equations:

$$v_d = i_d R_1 + d\lambda_d/dt + \lambda_q d\theta/dt \qquad 7.3$$

$$v_q = i_q R_1 + d\lambda_q/dt - \lambda_d d\theta/dt \qquad 7.4$$

During the transient, the speed $d\theta/dt = \omega$ can often be assumed constant.

Before the transient, $i_d = i_q = 0$ (open-circuit), $\lambda_d = i_{fo}L_{md}$, and $\lambda_q = 0$, where i_{fo} = field current referred to stator.

Hence $v_d = 0$, $v_q = V_q = -\omega L_{mdo} i_{fo}$ = constant.

The effect of the short-circuit can be simulated by applying axis voltages $v_d = 0$ and $v_q = + \omega L_{md} i_{fo}$ as step functions with $e_f = 0$, solving for the currents, flux linkages and voltages and adding these to the initial conditions.

If we neglect all resistances, we shall obtain currents which are correct only immediately after the short-circuit, but the analysis is simplified.

The equations for the transient components (i.e. without the initial current i_{fo} and its d-axis flux linkage) therefore become

$$0 = d\lambda_d/dt + \omega\lambda_q \qquad 7.5$$

$$-V_q = -\omega\lambda_d + d\lambda_q/dt \qquad 7.6$$

The transformer-type voltage components are represented by the simplified form of Fig. 7.1 shown in Fig. 7.5.

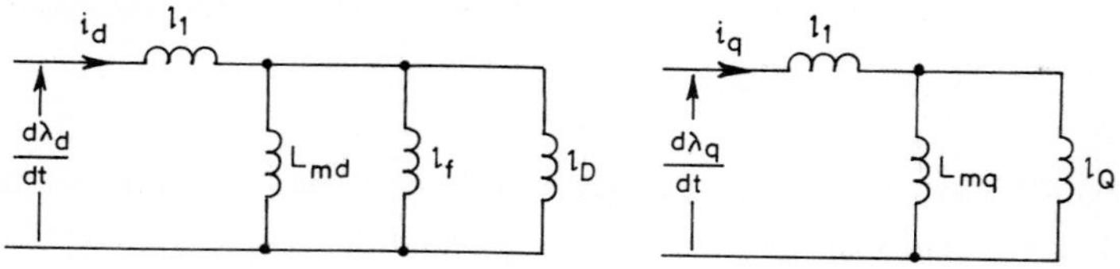

Fig. 7.5. Equivalent circuits for transformer-type components

These circuits are expressed by

$$\lambda_d = i_d l_d'' \text{ and } \lambda_q = i_q l_q'' \qquad 7.7$$

where $l_d'' = l_1 + L_{md}l_f l_D/(L_{md}l_f + L_{md}l_D + l_f l_d)$

$= \text{effective inductance for } \lambda_d$

and $l_q'' = l_q + L_{mq}l_Q/(L_{mq} + l_q)$

$= \text{effective inductance for } \lambda_q$

In terms of these parameters, eqns. 7.5 and 7.6 become

$$0 = x_d''di_d/d(\omega t) + x_q''i_q \qquad 7.8$$

$$V_q = x_d''i_d - x_q''di_q/d/(\omega t)$$

where $x_d'' = \omega l_d''$ and $x_q'' = \omega l_q''$ 7.9

From eqns.7.8 and 7.9,

$$d^2i_d/d(\omega t)^2 + i_d = V_q/x_d'' \qquad 7.10$$

and $i_q = -(x_d''/x_q'')di_d/d(\omega t)$ 7.11

Eqns. 7.10 and 7.11 solve, with $i_d = 0$ and $i_q = 0$ at $t = 0$, to give

$$\begin{aligned} i_d &= (V_q/x_d'')(1 - \cos\omega t) \\ &= (V_q/x_d'') - (V_q/x_d'')\cos\omega t \end{aligned} \qquad 7.12$$

$$i_q = -(V_q/x_q'')\sin\omega t \qquad 7.13$$

Those terms in eqns. 7.12 and 7.13 varying with time transform to unidirectional and second-harmonic components of phase currents (the latter usually being small) and the term V_q/x_d'' transforms to alternating phase currents at rotational frequency. These currents correspond to the "initial" values of d.c. and a.c. components identified in the recordings previously discussed.

The reactances x_d" (=ωl_d") and x_q" (=ωl_q") are the short-circuit sub-transient reactances.

If all resistances were zero, the current calculated above would persist indefinitely. In practice, resistances cause an energy loss which results in decay of the currents.

A more rigorous analysis [21] shows that the alternating d- and q- axis currents (i.e. the d.c. phase currents) decay with a time constant T_a which is approximately:

$$T_a = 2l_d''l_q''/(l_d'' + l_q'')R_1 \qquad 7.14$$

This is the short-circuit armature d.c. time constant.

The unidirectional axis current, only existing in the d-axis and corresponding to the sub-transient component of alternating phase current, decays at a rate determined by the resistance of the rotor circuits.

The d-axis equivalent circuit, following the short-circuit, is shown in Fig. 7.6 where i_d has an initial, unidirectional value V_q/x_d".

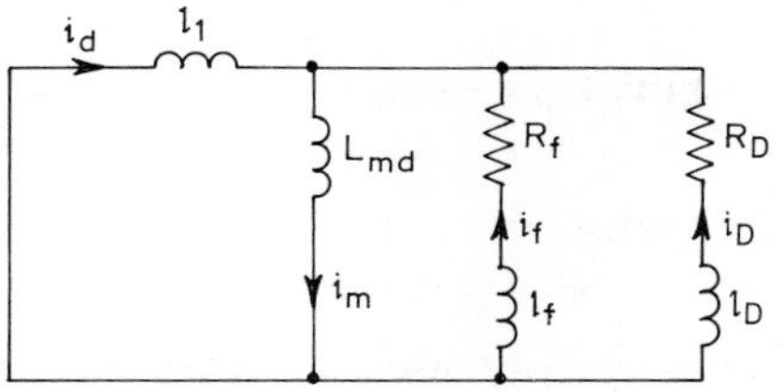

Fig. 7.6. Equivalent circuit on d–axis following short–circuit

The initial value of damper current is

$$i_D = -(V_q/x_d'')\ [L_{md}l_f/(L_{md}l_D + l_fl_D + l_fL_{md})]$$

and corresponds to the sudden generation of d–axis damper mmf.

Now usually the damper circuits have the highest resistance (referred to the stator), so the current most quickly affected by circuit resistance will be i_D. If $R_D >> R_f$, then i_D will decay much faster than i_f, but in doing so will affect all other currents. The effective inductance "seen" by i_D is $[l_D + L_{md}l_f/(L_{md}l_f + l_fl_1 + L_{md}l_1]$, that is, l_D in series with L_{md}, l_1 and l_f in parallel. So i_D will decay with a time constant T_d'' equal to

$$T_d'' = [l_D + L_{md}l_fl_1/(L_{md}l_f + l_fl_1 + L_{md}l_1)]/R_D \qquad 7.15$$

The decay of i_D causes the fall in i_D, and hence a.c. phase currents, and the rise in field current during the sub-transient period.

When i_D has decayed, i_f decays with a time constant T_d' given by

$$T_d' = [l_f + L_{md}l_1/(L_{md} + l_1)]/R_f \qquad 7.16$$

during the transient period.

If the damper circuits were excluded from the above treatment, the calculated currents would have initial values of:

$$i_d = (V_q/x_d') - (V_q/x_d')\cos\omega t$$

$$i_q = -(V_q/x_q')\sin\omega t$$

Here $x_d' = wl_d'$, and $x_q' = \omega l_q'$ are the short-circuit transient reactances, with

$$l_d' = l_1 + L_{md}l_f/(L_{md} + l_f) \qquad 7.17$$

$$l_q' = l_1 + L_{mq} = L_q\text{, the q-axis synchronous reactance} \qquad 7.18$$

This would give a transient phase-current variation similar to that discussed above, but without the sub-transient period.

For some machine transient problems, especially when the machine is treated, with others, as a component of a power system, the sub-transient is ignored.

The equivalent circuits and equations given above can be useful for analysing many types of transient problems such as the effects of switching or reswitching a machine to a system, including the starting condition, or system faults.

The equivalent circuits are also of some use in helping to interpret analytical solutions to transient problems, or observed behaviour.

For a simple example, consider the effects of a change in field voltage, such as might occur with a fast-acting voltage regulator, with the stator open-circuited. Effects are confined to the d-axis, and the equivalent circuit referred to the stator becomes as shown in Fig. 7.7, with $v_d = d\lambda_d/dt$ and $v_q = -\omega\lambda_d$.

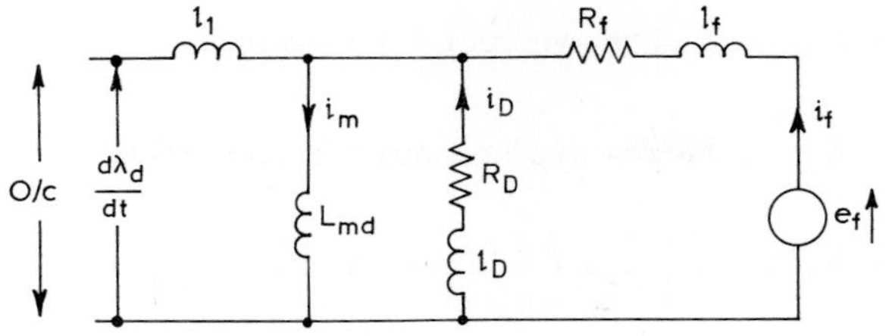

Fig. 7.7. Equivalent circuit for change in field voltage

The final steady-state condition will be when $i_f = e_f/R_f$, $i_D = 0$, $i_m = i_f$ and $\lambda_d = i_f L_{md}$, giving $v_d = 0$, $v_q = -\omega L_{md} i_f$, corresponding to the steady open-circuit voltage of the stator. During the transient, however, the damper circuits will influence the changes of i_f and λ_d. Initially, e_f will see a low-impedance circuit owing to the coupled damper circuit. Assuming $L_{md} >> l_D$ or l_f, then i_f and $-i_D$ will rapidly reach a high value of approximately $e_f/(R_f + R_D)$, and i_m will be very low.

Then i_D will decay rapidy, if $R_D >> R_f$ which is usually the case, with a time constant:

$$T_{do}'' = [l_D + l_f L_{md}/(l_f + L_{md})]/R_D \qquad 7.19$$

and i_m, and hence λ_d, will increase in sympathy. This is the open-circuit sub-transient period. Thereafter $i_f \simeq i_{md}$, and increases with time constant:

$$T_{do}'' = (l_f + L_{md})/R_f \qquad 7.20$$

during the open-circuit transient period.

7.4 Summary of Machine Constants

Some combinations of a.c. machine parameters are of use in many analyses, and the most commonly used are listed below.

Note that all quantities are referred to the stator per phase.

$X_d = x_1 + X_{md}$ = d–axis synchronous reactance

$X_q = x_1 + X_{mq}$ = q–axis synchronous reactance

$x_d' = x_1 + X_{md}x_f/(X_{md} + x_f)$ = d–axis transient reactance

$x_d'' = x_1 + X_{md}x_fx_D/(X_{md}x_f + X_{md}x_D + x_fx_D)$ = d–axis sub-transient reactance

$x_q'' = x_1 + X_{mq}x_Q/(X_{mq} + x_Q)$ = q–axis sub-transient reactance

$x_2 = (x_d'' + x_q'')/2$ = negative phase sequence reactance (provided that x_d'' and x_q'' do not differ greatly)

$T_{do}' = [l_f + L_{md}]/R_f$ = d–axis, open–circuit, transient time constant

$T_d' = [l_f + l_{md}l_1/(L_{md} + l_1)]/R_f$ = d–axis, short–circuit, transient time constant

$T_{do}'' = [l_D + L_{md}l_f/(L_{md} + l_1)]/R_D$ = d–axis, open–circuit, sub-transient time constant

$T_d'' = [l_D + L_{md}l_fl_1/(L_{md}l_f + L_{md}l_1 + l_fl_1)]/R_D$ = d–axis, short–circuit sub-transient time constant

$T_{qo}'' = [l_Q + L_{mq}]/R_Q$ = q–axis, open–circuit sub-transient time constant

$T_q'' = [l_Q + L_{mq}l_1/(L_{mq} + l_1)]/R_Q$ = q–axis, short–circuit, sub-transient time constant

$T_D = l_Q/R_Q$ = d–axis damper leakage time constant

$T_a = 2l_d''l_q''/(l_d'' + l_q'')R_1$ = stator, short–circuit, time constant.

7.5 Ranges of Typical Values for Large Machines

Reactances, per-unit

x_1	0.1	-	0.25
X_d	0.5	-	2.5
X_q	0.35	-	2.5
x_d'	0.2	-	0.35
x_d''	0.1	-	0.3
x_q''	0.1	-	0.8

Time constants, seconds

T_d'	1.0	-	2.5
T_d''	0.03	-	0.1
T_a	0.1	-	0.2

7.6 Induction-Motor Transients

It should be mentioned that the analysis of induction-motor transients requires a treatment similar to that used for synchronous machines. The major difference is that induction motors are usually symmetrical so that the d-axis and q-axis parameters are equal, and winding currents are consequently slightly less complex in their behaviour following transients.

The axis equivalent circuits are then identical and of the form shown in Fig. 7.8.

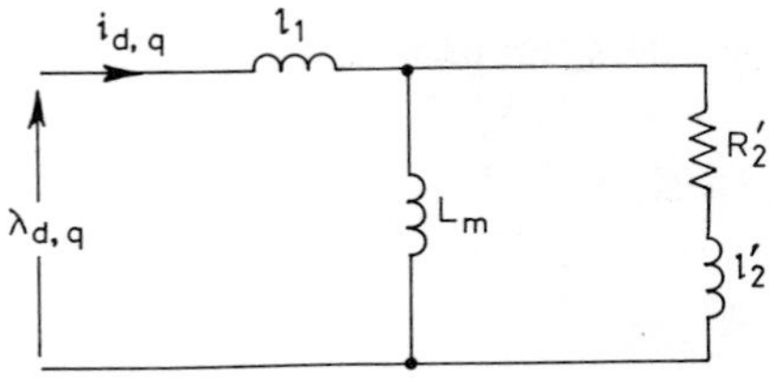

Fig. 7.8. Axis equivalent circuits for induction-motor transients

Care should be taken to ensure that R_2' and l_2' are calculated for an appropriate rotor frequency if skin effect is pronounced.

7.7 Calculation of Synchronous-Machine Parameters

7.7.1 Stator Quantities

Stator leakage reactance.

x_1 can be calculated by the method given in Section 4.2.

d-axis magnetising reactance X_{md}

The mmf developed by a three-phase current I, r.m.s. , is

$$F_1 = (3/2)(4/\pi)(NK_{w1}/2p)\sqrt{2}I$$

If this mmf is acting on the d-axis, the amplitude of the fundamental component of flux density has a value, from Section 3.2 , of:

$$B_1 = A_{d1}(\mu_o/g_{mine})F_1$$

corresponding to a flux per pole Φ, given by

$$\Phi_1 = (2/\pi)\tau LB_1 = (2/\pi)\tau(\mu_o/g_{mine})LA_{d1}F_1$$

$$= \lambda_a LA_{d1}F_1$$

and inducing a voltage, r.m.s.per phase, of

$$E = (2\pi f/\sqrt{2})NK_{w1}\Phi_1$$

$$= (\omega/\sqrt{2})NK_{w1}\Phi_1$$

Now $E = X_{md}I = \omega L_{md}I$

Substituting gives

$$L_{md} = (3/\pi)(NK_{w1}/p)^2 L\lambda_a A_{d1}$$

and $$X_{md} = \omega L_{md} \qquad 7.21$$

q-axis magnetising reactance X_{mq}

If the mmf acted on the q-axis, rather than the d-axis, the fundamental flux density would be

$$B_1 = A_{q1}(\mu_o/g_{mine})F_1$$

Hence, by similar deduction,

$$L_{mq} = (3/\pi)(NK_{w1}/p^2)L\lambda_a A_{q1}$$

and $$X_{mq} = \omega L_{mq} \qquad 7.22$$

7.7.2 Referred Field Quantities

The field has circuits only on the d-axis, so first consider the field referred to ONE stator phase winding. The mmf produced by a current i_f' in one phase of the stator only will be

$$(4/\pi)(NK_{w1}/2p)i_f'$$

equivalent to the field mmf $i_f N_f$.

We must equate the fundamental components of flux densities which would be produced by these mmfs if they acted alone. Hence

$$(4/\pi)(NK_{w1}/2p)A_{d1}i_f' = i_f N_f A_1$$

giving stator:field turns ratio :

$$i_f/i_f' = (2/\pi)(NK_{w1}/pN_f)(A_{d1}/A_1) \qquad 7.23$$

A field mmf i_fN_f produces a fundamental flux per pole

$$\Phi_1 = (2/\pi)\tau(L\mu_o/g_{mine})A_1N_fi_f$$

$$= \lambda_aLA_1N_fi_f$$

This gives a flux linkage with <u>one</u> phase of the stator given by

$$M_{af}i_f = NK_{w1}\Phi_1.$$

Hence

$$M_{af} = \lambda_aLA_1N_fK_{w1} = \text{mutual inductance} \qquad 7.24$$

The total inductance of the field winding is given by the flux linkage per ampere:

$$L_{ff} = N_f(\Phi_{TOT}/i_f)2p \qquad 7.25$$

where Φ_{TOT} is total flux linking the field winding of one pole produced by mmf i_fN_f.

Section 3.2.1 showed that the mmf i_fN_f produces a flux crossing the airgap which is K_φ times the fundamental flux Φ_1, and this total airgap flux can be written as $K_\varphi\lambda_aLA_1N_fi_f$

In addition to this, flux is produced which does not reach the stator and could be written as $\lambda_fLN_fi_f$ where λ_f is a leakage permeance per unit length. This includes pole-end leakage as well as the pole-tip and pole-body leakage discussed in Section 3.2.1. Expressions for the permeance components of λ_f can be found in References [7] and [8].

Hence $\quad \Phi_{TOT} = \lambda_aLN_fi_f[K_\varphi A_1 + \lambda_f/\lambda_a] \qquad 7.26$

and, for 2p poles,

$$L_{ff} = \lambda_a L N_f^2 \; [K\varphi A_1 + \lambda_f/\lambda_a]2p \qquad 7.27$$

It was shown in Section 1.5 that for two coils an equivalent circuit referred to coil 1 can be drawn on the basis of a turns ratio N_1/N_2, so that

$$l_2' = L_2(N_1/N_2)^2 - (N_1/N_2)M$$

where l_2' is referred value of coil-2 leakage inductance, L_2 = coil-2 self-inductance and M is the mutual inductance between them.

The following relationships derived above are used:

$$(N_1/N_2) \equiv (2/\pi)(NK_{w1}/pN_f)(A_{d1}/A_1)$$

$$L_2 \equiv \lambda_a L N f^2 \; [K_\varphi A_1 + (\lambda_f/\lambda_a)]2p = L_{ff}$$

$$M \equiv \lambda_a L A_1 N_f N K_{w1} = M_{af}$$

These give

$$(N_1/N_2) \; M \equiv (2/3)(3/\pi)(NK_{w1}/p)^2 \; L\lambda_a A_{d1} \qquad 7.28$$

which will be recognised as $(2/3)L_{md}$

and

$$(N_1/N_2)^2 L_2 \equiv (2/3)L_{md}(4/\pi)(A_{d1}/A_1^2)[K\varphi A_1 + (\lambda_f/\lambda_a)] \qquad 7.29$$

Hence $\quad l_2' = (2/3)L_{md}[(4/\pi)(A_{d1}/A_1)(K\varphi + \lambda_f/A_1\lambda_a)-1] \qquad 7.30$

This is the field leakage inductance referred to ONE stator winding phase.

For a three-phase winding, the flux linkage per phase produced by flux crossing the airgap is 3/2 times that which would occur for only a single phase carrying current. Since all referred rotor quantities are experienced by the stator through the coupling effect in the airgap, then all referred parameters will be increased in per-phase value by 3/2 compared with those seen by a single-phase winding.

Thus, the quantity represented by $(N_1/N_2)M$ above becomes $(3/2)(N_1/N_2)M$ per phase, which is equal to L_{md} as expected (see eqn. 7.21). The referred field leakage inductance per phase l_f is:

$$l_f = L_{md}\ [(4/\pi)(A_{d1}/A_1)\{K\varphi + (\lambda_f/\lambda_a A_1)\}-1] \qquad 7.31$$

The field resistance can be obtained as $(3/2)(N_1/N_2)^2 R_{field}$

whence

$$R_f = (6/\pi^2)[NK_{w1}A_{d1}/pN_fA_1]^2 R_{field} \qquad 7.32$$

It should be noted that the value above is valid for steady-state operation. For example, Section 3.2.2 derived a current I_e which is an r.m.s. 3-phase current equivalent to the field current I_f, and such that stator induced r.m.s. voltage per phase is given by

$$E = \omega L_{md} I_e$$

with $\quad I_e = (\pi/3\sqrt{2})(A_1/A_{d1})(pN_f/NK_{w1})I_f$

One would expect, therefore, that ohmic loss should be the same in actual and referred terms.

That is,

$$3I_e^2 R_f = I_f^2 R_{field} \qquad 7.33$$

Substitution for I_e and R_f shows that this is indeed the case.

7.7.3 Referred Damper Quantities

Damper circuits can be of the form of bars in slots on the pole face, shorted together at the pole-ends, and with or without interconnections between poles. Alternatively the damping may be derived from currents flowing in the iron of solid pole faces, with or without interconnectors between poles, or even a combination of solid poles and damper bars. These latter possibilities are difficult to analyse and no completely satisfactory solutions has been published, so that machine designers depend upon empirical methods for estimating quantities.

Even for the case of bars in slots, in a laminated pole system, the problem is complex. Consider the damper system shown in Fig. 7.9.

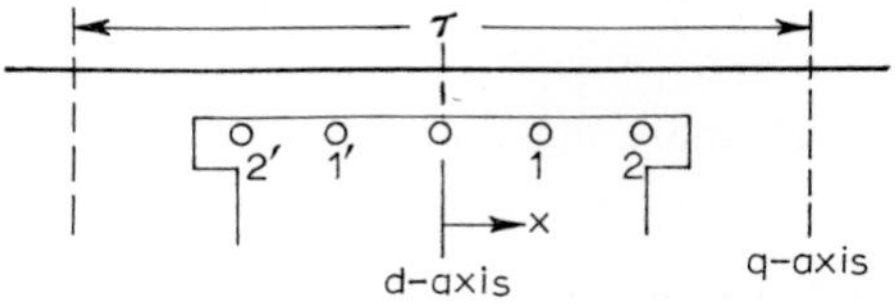

Fig. 7.9. Distribution of damper bars

The d-axis stator mmf has a value at distance x from the d-axis of

$$F_1\cos(x\pi/\tau) \qquad 7.34$$

Only if this is changing will currents be induced in the damper bars, but the problem is to determine the rotor-current distribution. An exact solution would require the treatment of every damper bar (or pair of bars such as 1, 1' or 2, 2') separately, with account being taken of the coupling between all pairs of bars and the stator winding in each case. Some attempts have been made to do this [22, 23, 24] but no general expression which is convenient for design-office use has been produced.

An approximation is to assume that the damper-bar currents have a sinusoidal distribution about the axis centre-line. This approach was used by Talaat [25] to produce a reasonably convenient set of formulae, which appear to give satisfactory agreement with test results.

The formulae only apply for the case of uniform spacing of identical damper bars and with interpolar connections between short-circuit rings.

The d-axis and q-axis equivalent damper leakage inductances, referred to the stator, are:

$$l_D = L_{md}(2\pi/A_{d1})(\lambda_{bed}/n_b\lambda_a)/(1 - k_b) \qquad 7.35$$

$$l_Q = L_{md}(2\pi/A_{d1})(\lambda_{beq}/n_b\lambda_a)/(1 + k_b) \qquad 7.36$$

where n_b = number of damper bars per pole

α_b = electrical angle between bars

$k_b = \sin(n_b\alpha_b)/n_b\sin\alpha_b$

and λ_{bed} (or λ_{beq}) is the leakage permeance per unit length of a single bar. The difference between λ_{bed} and λ_{beq} arises from any allowance made for end-ring effects but such an allowance could be neglected.

The d-axis and q-axis equivalent damper resistances referred to the stator are

$$R_D = 6\{(NK_{w1})^2/p\}(r_{bed}/n_b)/(1 - k_b), \qquad 7.37$$

$$R_Q = 6\{(NK_{w1})^2/p\}(r_{beq}/n_b)/(1 + k_b), \qquad 7.38$$

where r_{bed} (or r_{beq}) is the resistance of a single bar, and the difference between r_{bed} and r_{beq} again arises from the allowance made for end-effect. For an approximate estimate such an allowance could be based on an approach similar to that used for cage induction motors.

Plate 7. Salient-pole rotor of a 6MW, 300 rev/min synchronous motor to drive a log grinder in a pulp mill, with alloy pole-face cage winding. Starting current = 4.5 times full-load current, starting torque = 0.75 times full-load torque. Note the expansion joints on the interpolar connections, to remove stress from the cage bars

Courtesy of Peebles Electrical Machines

CHAPTER 8
References

[1] Jayawant, B.V.: 'Induction Machines', McGraw-Hill, 1968, p.43

[2] Burbidge, R.F.: Proc. IEE, 105C, 1958, p.307.

[3] Chalmers, B.J.: Proc. IEE, 111, 1964, p.1859.

[4] Carter, F.W.: Journ. IEE, 64, 1926, p.1115.

[5] Binns, K.J.: Proc. IEE, 111, 1964, p.1847.

[6] Wieseman, R.W.: Trans. AIEE, 46, 1927, p.141.

[7] Doherty, R.E. and Shirley, O.E.: Trans. AIEE, 37, 1918, p.1209.

[8] Kilgore, L.A.: Trans. AIEE, 50, 1931, p.1201.

[9] Lawrenson, P.J. and Gupta, S.K.: Proc. IEE, 114, 1967, p.645.

[10] Fong, W. and Hisui, J.S.C.: Proc. IEE, 117, 1970, p.545.

[11] Chalmers, B.J. and Mulki, A.S.: IEEE Trans., PAS-91, 4, 1972, p. 1562.

[12] Cruickshank, A.J.D., Menzies, R.W. and Anderson, A.F.: Proc. IEE, 113, 1966, p.2058.

[13] Alger, P.L.: 'The Nature of Induction Machines', Gordon and Breach, 1965, p.203.

[14] ibid., p. 216.

[15] ibid., p.227.

[16] Butler, O.I. and Birch, T.S.: Proc. IEE, 118, 1971, p.879.

[17] Chalmers, B.J. and Dodgson, R.: Proc. IEE, 116, 1969, p.1395.

[18] Carter, G.W.: 'The Electromagnetic Field in its Engineering Aspects', Longmans, 1954, p.244.

[19] Alger, P.L.: loc. cit., p.265.

[20] Bruges, W.E.: Proc. Roy. Soc. Edinburgh, 62 II, 1946, p.175.

[21] Adkins, B.: 'The General Theory of Electrical Machines', Chapman and Hall, 1959, p.147.

[22] Linville, T.M.: Trans. AIEE, 49, 1930, p.531.

[23] Rankin, A.W.: AIEE Trans., 64, 1945, p.861.

[24] Menon, K.B.: AIEE Trans., 78, 1959, p.371.

[25] Talaat, M.E.: AIEE Trans., 74, 1955, p.176, and 75, 1956, p.317.

APPENDIX A
Units and Dimensions

Quantity	Unit name	Unit symbol	Unit dimensions	Derivation
Base Units:				
length, L	metre	m	m	
mass, M	kilogram	kg	kg	
time, t	second	s	s	
electric current, I	ampere	A	A	
Derived units:				
velocity, v		m/s	$m.s^{-1}$	
acceleration, a		m/s^2	$m.s^{-2}$	
angle, θ	radian	rad	–	
angular velocity, ω		rad/s	s^{-1}	
angular acceleration, α		rad/s^2	s^{-2}	
frequency, f	hertz	Hz	s^{-1}	
force, F	newton	N	$m.kg.s^{-2}$	F=Ma
torque, T		Nm	$m^2.kg.s^{-2}$	T=FL
moment of inertia, J		$kg.m^2$	$m^2.kg$	

energy, E	joule	J	$m^2.kg.s^{-2}$	E=FL
power, P	watt	W	$m^2.kg.s^{-3}$	P=E/t
electric charge, Q	coulomb	C	s.A	Q=It
electric potential, V	volt	V	$m^2.kg.s^{-3}.A^{-1}$	V=P/I
capacitance, C	farad	F	$m^{-2}.kg^{-1}.s^4.A^2$	C=Q/V
resistance, R	ohm	Ω	$m^2.kg.s^{-3}.A^{-2}$	$P=I^2R$
magnetic flux, Φ	weber	Wb	$m^2.kg.s^{-2}.A^{-1}$	V=dΦ/dt
magnetic flux density, B	tesla	T	$kg.s^{-2}.A^{-1}$	$B=\Phi/L^2$
inductance, L	henry	H	$m^2.kg.s^{-2}.A^{-2}$	$E=\frac{1}{2}LI^2$
magnetomotive force, F	ampere	A	A	
magnetising field, H		A/m	$m^{-1}.A$	I=HL
current density, J		A/m^2	$m^{-2}.A$	
permeability, μ	–	H/m	$m.kg.s^{-2}.A^{-2}$	$B=\mu H$
permittivity, ε	–	F/m	$m^{-3}.kg^{-1}.s^4.A^2$	$\mu_o\epsilon_o=1/v^2$

APPENDIX B
Constants and Conversion Factors

Constants

Permeability of free space	μ_o	=	$4\pi \times 10^{-7}$ H/m
Permeability of free space	ϵ_o	=	8.854×10^{-12} F/m
Acceleration of gravity	g	=	9.807 m/s^2

Conversion factors

Length	lm	=	3.281 ft
		=	39.37 in
Mass	lkg	=	0.0685 slug
		=	2.205 lb (mass)
Force	1 N	=	10^5 dyne
		=	0.2248 lbf
		=	7.233 poundals
Torque	1 Nm	=	0.738 lb ft
Energy	1 J	=	10^7 erg
		=	0.7376 ft lbf
		=	0.2388 cal (1 cal = 4.186 J)
		=	9.48×10^{-8} Btu
Power	1 W	=	(1/746)hp = 1.341×10^{-3} hp
Moment of inertia	1 kgm^2	=	0.738 slug ft^2
		=	23.7 lb ft^2

Magnetic flux	1 Wb	=	10^8 maxwell (lines)
Magnetic flux density	1 T	=	10^4 gauss
		=	64.52 kilolines/in^2
Magnetising force	1 A/m	=	0.0254 A/in
		=	0.01257 oersted

APPENDIX C
Additional References

Harris, M.R. Lawrenson, P.J. and Stephenson, J.M.: Per-unit systems, with special reference to electrical machines, Cambridge University Press, 1970.

Canay, M.: Asynchronous starting of synchronous machines with or without rectifiers in the field circuit, Proc. IEE, 119(12), pp. 1701–1708, 1972.

Smith, J.R., Binns, K.J., Williamson, S. and Buckley, G.W.: Determination of saturated reactances of turbogenerators, Proc. IEE, 137(3), Pt. C. pp. 122–128, 1980.

Reece, A.B.J., Khan, G.K.M. and Chant, M.J.: Generator parameter prediction from computer simulation of standstill variable-frequency injection test, IEE Conf. on Electrical Machines: Design and Applications, Conf. Pubn. 213, pp. 99–103, 1982.

Preston, T.W. and Reece, A.B.J.: The contribution of the finite element method to the design of electrical machines: an industrial viewpoint, IEEE Trans. on Magnetics, 19, pp. 2375–2380, 1983.

Macdonald, D.C., Reece, A.B.J. and Turner, P.J.: Turbine-generator steady-state reactances, IEE Proc. C, 132(3), pp. 101–108, 1985.

Turner, P.J.: Finite-element simulation of turbine-generator terminal faults and application to machine parameter prediction, IEEE Trans. EC-2(1), pp. 122-131, 1987.

Williamson, S. and Lloyd M.R.: Cage rotor heating at standstill, IEE Proc. 134, B, pp.325-332, 1987.

Preston, T.W., Reece, A.B.J. and Sangha, P.S.: Induction motor analysis by time-stepping techniques, IEEE Trans. on Magnetics, 24, pp. 471-474, 1987.

Harris, M.R. and Prashad, F.R.: Improved methods for relating equivalent circuits and frequency response data for synchronous machines, IEE Int. Conf. on Electrical Machines and Drives, Conf. Pubn. 310, pp. 192-197, 1989.

Chalmers, B.J. (Ed.): Electric Motor Handbook, Butterworths, 1988.

De Jong, H.C.J.: A.C. Motor design: rotating magnetic fields in a changing environment, Hemisphere Publishing Corporation , 1989.

Smith, J.R.: Response analysis of A.C. electrical machines, Research Studies Press, 1990.

Index